Electronic Devices and Circuit Fundamentals

Solutions Manual

RIVER PUBLISHERS SERIES ELECTRONIC MATERIALS, CIRCUITS AND DEVICES

The "River Publishers Series in Electronic Materials, Circuits and Devices" is a series of comprehensive academic and professional books which focus on theory and applications of advanced electronic materials, circuits and devices. This includes analog and digital integrated circuits, memory technologies, system-on-chip and processor design. Also theory and modeling of devices, performance and reliability of electron and ion integrated circuit devices and interconnects, insulators, metals, organic materials, micro-plasmas, semiconductors, quantum-effect structures, vacuum devices, and emerging materials. The series also includes books on electronic design automation and design methodology, as well as computer aided design tools.

Books published in the series include research monographs, edited volumes, handbooks and textbooks. The books provide professionals, researchers, educators, and advanced students in the field with an invaluable insight into the latest research and developments.

Topics covered in this series include:-

- Analog Integrated Circuits
- Data Converters
- Digital Integrated Circuits
- Electronic Design Automation
- Insulators
- Integrated circuit devices
- Interconnects
- Memory Design
- MEMS
- Nanoelectronics
- Organic materials
- Power ICs
- Processor Architectures
- Quantum-effect structures
- Semiconductors
- Sensors and actuators
- System-on-Chip
- Vacuum devices

For a list of other books in this series, visit www.riverpublishers.com

Electronic Devices and Circuit Fundamentals Solutions Manual

Dale R. Patrick

USA

Stephen W. Fardo

Eastern Kentucky University, USA

Ray E. Richardson

Eastern Kentucky University, USA

Vigyan (Vigs) Chandra

Eastern Kentucky University, USA

NEW YORK AND LONDON

Published 2023 by River Publishers

River Publishers

Alsbjergvej 10, 9260 Gistrup, Denmark

www.riverpublishers.com

Distributed exclusively by Routledge

605 Third Avenue, New York, NY 10017, USA

4 Park Square, Milton Park, Abingdon, Oxon OX14 4RN

Electronic Devices and Circuit Fundamentals Solutions Manual / by Dale R. Patrick, Stephen W. Fardo, Ray E. Richardson, Vigyan (Vigs) Chandra.

Routledge is an imprint of the Taylor & Francis Group, an informa business

ISBN 978-87-7022-815-2 (print)

978-10-0089-699-2 (online)

978-1-003-40327-2 (ebook master)

While every effort is made to provide dependable information, the publisher, authors, and editors cannot be held responsible for any errors or omissions.

Preface

Electronics: Devices and Circuit Fundamentals by Patrick, Fardo, Richardson and Chandra explores many fundamental topics in a basic and easy-to-understand manner. This book, and the accompanying **DC-AC Electrical Fundamentals** by the same co-authors, have been developed using a classic textbook– **Electricity and Electronics: A Survey (5th Edition)** by Patrick and Fardo– as a framework. Both new books have been structured using the same basic sequence and organization of the textbook as previous editions. The previous edition of **Electricity and Electronics: A Survey** contained 18 chapters, 8 in the **Electricity** section and 10 in the **Electronics** section.

Electronics: Devices and Circuit Fundamentals has been expanded to include 22 chapters, further simplifying content and providing a more comprehensive coverage of fundamental content. Expanded content for this textbook includes 26 chapters. The content has been continually updated and revised through new editions and by external reviewers throughout the years. Additional quality checks to insure technical accuracy, clarity and coverage of content have always been an area of focus. Each edition of the text has been improved through the following features:

1. Improved and updated text content
2. Improved usage of illustrations and photos
3. Use of color to add emphasis and clarify content

Organization of the Solutions Manual (by Chapter)

- Chapter Outline
- Learning Objectives
- Key Terms
- Figure List
- Chapter Summary
- Formulas

- Answers to Examples / Self-Exams
- Glossary of Terms (defined)

Dale R. Patrick
Stephen W. Fardo
Ray E. Richardson
Vigyan (Vigs) Chandra

About the Authors

Dale Patrick was a Professor in the Department of Applied Engineering and Technology at Eastern Kentucky University. His experience includes teaching and organizing laboratory classes in Electrical Engineering/Electronics Technology for many years. He completed Bachelor of Science and Graduate Degrees at Indiana State University. His experience also includes research projects, technical teacher training, and energy consulting for business and industry. He and Dr. Fardohave co-authored over 30 textbooks and laboratory manuals.

Stephen W. Fardois currently a Foundation Professor in the Department of Applied Engineering and Technology at Eastern Kentucky University. His experience includes teaching and organizing laboratory classes in Electrical Engineering/Electronics Technology for many years and coordinating/advising technical teacher programs for the State of Kentucky. He completed Bachelor of Science, Master of Science and Doctoral degrees at Eastern Kentucky University and University of Kentucky. His experience also includes research projects, technical teacher education, and energy consulting for business and industry. He and Dale Patrick have co-authored over 30 textbooks and laboratory manuals.

Ray E. Richardson is a Professor in the Department of Applied Engineering and Technology at Eastern Kentucky University. His experience includes teaching and organizing lecture and laboratory classes in Electrical Engineering/Electronics Technology since 1985 as well as teaching technical seminars, technical writing, and classes in the graduate program. He completed Bachelor of Science and Master of Science degrees at Eastern Illinois University, and a doctorate level degree at the University of Illinois. He is a certified Technology Manager at Senior level by ATMAE. Other experience includes research projects in technical education and consulting in the manufacturing, food, and electronics areas.

Vigyan (Vigs) Chandra is a professor and coordinator of the Cyber Systems Technology-related undergraduate and graduate degree programs offered by the Departments of Computer Science and Information Technology (CSIT) and Applied Engineering and Technology (AET) at Eastern Kentucky University. He earned a doctoral degree from the University of Kentucky in Electrical and Computer Engineering and a master's degree in Career and Technical Education from Eastern. He holds certifications in various computer networking areas and teaches computer systems and applications, network hardware, communication systems, and digital, analog, and machine-control electronics. He is the recipient of the 2020–21 Critical Thinking and Reading Teacher Awards and the 2013 Golden Apple Award for Teaching Excellence at Eastern. His professional interests include implementing active teaching and learning strategies, metacognition, and integrating open-source software/hardware with online control and computer networking technologies.

Electronic Devices and Circuits

Section I – Electronic Devices

Chapter 1: Semiconductor Fundamentals – Chapter Outline

Chapter 1 Objectives

After studying this chapter, you will be able to:

1.1 Describe the structure and properties of an atom.
1.2 Compare and contrast ionic bonding, covalent bonding, and metallic bonding.

1

1.3 Explain the electrical difference between conductors, semiconductors, and insulators.

1.4 Explain how current carriers move through semiconductors.

1.5 Analyze and troubleshoot semiconductors.

Chapter 1 Key Terms

atom
atomic number
atomic weight
breakdown voltage
compound
covalent bonding
doping
electron
electron volt
element
extrinsic material
forbidden gap
intrinsic material
ionic bonding
matter
metallic bonding
molecule
neutron
N-type material
nucleus
orbital
proton
P-type material
valence electrons

Chapter 1 – Semiconductor Fundamentals – Figure List

Figure 1-1. The relationship of matter, elements, and atoms. Matter can consist of one or more types elements. An element consists of the same type of atoms.

Figure 1-2. Water formed by combining hydrogen and oxygen; (a) hydrogen atoms; (b) oxygen atom; (c) water molecule.

Figure 1-3. Two-dimensional model of a copper atom. The nucleus of an atom contains protons and neutrons. Electrons orbit the nucleus.

Figure 1-4. Periodic table of the elements. a) The periodic table, b) elements in the periodic table showing groups, symbols, atomic number, atomic weight and electron arrangement in shells 1 through 7 (K through Q).

Figure 1-5. Electrons orbit the nucleus in shells. A shell may have from one to six distinct energy levels in its structure.

Figure 1-6. Shells and energy levels of a copper atom. Shells K through M follow the typical electron pattern. The N-shell, however, does not contain the full number of electrons for its energy level.

Figure 1-7. Two-dimensional representation of a lithium atom. The K-shell follows the typical electron pattern. The L-shell does not.

Figure 1-8. Orbitals of an atom.

Figure 1-9. Ionic bonding of chlorine (CI) and sodium (Na) atoms to produce salt. When a valence electron of a chlorine is transferred to the M-shell of a sodium atom, chlorine becomes a negatively charged ion, while the sodium atom becomes a positively charged ion. The positive sodium (Na^+) and negative chlorine (CI^-) ions are bonded together by an electrostatic force.

Figure 1-10. Covalent bonding of hydrogen atoms. Electrons are shared in this type of bonding.

Figure 1-11. Metallic bonding of copper. In this type of bonding, electrons "float" around in a cloud that covers the positive ions. This "floating" cloud bonds the electrons randomly to the ions.

Figure 1-12. Energy-level diagrams for insulators, semiconductors, and conductors.

Figure 1-13. Intrinsic crystal of silicon. The electrons of the individual atoms are covalently bonded together.

Figure 1-14. Current carrier movement in silicon. Electrons and holes move in opposite directions.

Figure 1-15. N-type crystal material. The extra electron of each impurity atom does not take part in a covalent bonding group, and therefore, does not alter the crystal structure or bonding process.

Figure 1-16. Current carriers in an N-type material. Electrons are the majority current carriers and holes are the minority current carriers.

Figure 1-17. P-type crystal material. The atoms of the crystal material form a covalent bond with indium atoms, creating a deficiency or hole in the covalent bonding structure. This hole represents a positively charged area.

Figure 1-18. Current carriers in a P-type material. Holes are the majority current carriers and electrons are the minority current carriers.

Chapter 1 Summary

- An atom is the smallest particle to which an element can be reduced and still retain its identity.
- Atoms consist of smaller particles called *electrons*, *neutrons*, and *protons*.
- The nucleus (core) of every atom is composed of one or more positively charged particles called *protons* and one or more particles with no electrical charge called *neutrons*.
- For every proton in the nucleus of an atom there is a negatively charged particle called an *electron* that orbits the nucleus.
- Electrons orbit the nucleus in shells or layers.
- The energy level and location of electrons in the structure of an atom determines its electrical conductivity.
- Valence electrons are electrons in the outermost shell and represent the highest energy level of the atom.
- Stabilized atoms have a full complement of electrons in the valence band.
- A stabilized atom will not release electrons under normal conditions.
- Unstable atoms are those that do not have a full complement of electrons in the valence band.
- An unstable atom will try to become stable by drawing electrons away from neighboring atoms.
- Ionic bonding is the process of bonding atoms through electrostatic force.
- Electrostatic force occurs through the attraction of opposite net charges between two atoms.
- In covalent bonding, electrons alternately shift back and forth between each atom and this force bonds individual atoms in a simulated condition of stability.
- In metallic bonding, valence electrons of metal wander between different atoms and form a floating cloud of electrons that permits the material to be a good conductor.
- A specific amount of outside energy must be added to valence electrons to cause them to go into conduction.
- Conductivity is the basis of material classifications such as insulators, semiconductors, and conductors.
- Hole flow and electron flow occur in semiconductor material as long as energy is supplied.
- An N-type material is formed when intrinsic silicon is mixed with a Group VA element, such as arsenic.
- A P-type material is formed when intrinsic silicon is mixed with Group IIIA elements, such as indium or gallium.

- In contrast to intrinsic silicon crystal, extrinsic silicon crystal will go into conduction with a very small amount of applied voltage.
- In N-type material, electrons are the majority current carriers and holes are the minority carriers. In P-type material, holes are the majority current carriers and electrons are the minority carriers.

Chapter 1 Formulas

(1-1) atomic weight = atomic mass rounded to nearest whole number Atomic weight of an atom.

(1-2) number of neutrons = atomic weight – atomic number Number of neutrons in the nucleus of an atom.

Chapter 1 Answers

Examples

1-1. 14

1-2. K-shell = 2 s-level

L-shell = 2 s-level

= 6 p-level

M-shell = 2 s-level

= 6 p-level

= 10 d-level

N-shell = 2 s-level

= 2 p-level

Self-Examination

1.1

1. atomic numbe
2. protons, neutrons
3. mass
4. electron
5. K
6. s, p, d, e, f, and g
7. P
8. orbital

1.2

9. ground
10. stable
11. compound
12. ionic
13. ionic
14. covalently
15. covalent
16. Metallic

1.3

17. valence, conduction
18. valence
19. forbidden
20. electron volts
21. insulator
22. semiconductor
23. conductor
24. breakdown

1.4

25. intrinsic
26. conductive or a conductor
27. negative, positive
28. acceptor
29. donor
30. d. Four

Chapter 1 Glossary

matter
Anything that occupies space and has weight. Can be a solid, a liquid, or a gas.

element
The basic materials that make up all other materials. They exist by themselves, such as copper, hydrogen, and carbon or in combination with other elements, such as water, a combination of the elements hydrogen and oxygen.

atom
The smallest particle to which an element can be reduced and still retain its identity.

compound
Two or more elements that have been chemically combined.

molecule
The smallest particle to which a compound can be reduced before being broken down into its basic elements.

nucleus
The core or center part of an atom, which contains protons having a positive charge and neutrons having no electrical charge.

proton
A particle in the center of an atom that has a positive (+) electrical charge.

neutron
A particle in the nucleus (center) of an atom that has no electrical charge.

electron
A negatively charged particle that orbits the nucleus of an atom.

atomic weight
The sum of protons and neutrons in the nucleus.

atomic number
The number of protons contained in the nucleus of an atom.

valence electrons
Electrons in the outer shell of an atom.

orbital
The mathematical probability of where an electron will appear in the structure of an atom.

ionic bonding
Bonding that occurs between the attraction of opposite net charges (electro-static force) of two atoms.

covalent bonding
Bonding that occurs by atoms positioning themselves so that the energy levels of their valence electrons interact. The valance electrons, are in essence, shared between atoms.

metallic bonding
Bonding that occurs due to a floating cloud of ions that hold atoms loosely together in a conductor.

forbidden gap
Separates the valence and conduction bands. The width of the forbidden gap indicates the conduction status the material it represents. In atomic theory, the width of the forbidden gap is expressed in electron volts (eV).

electron volt
The amount of energy gained or lost when an electron is subjected to a potential difference of 1 V.

breakdown voltage
The energy level value that causes an insulator to go into conduction.

intrinsic material
A very pure semiconductor crystal.

doping
The process of adding an impurity to an intrinsic material.

extrinsic material
An intrinsic material that has been doped.

N-type material
A semiconductor material that is formed when intrinsic silicon is mixed with a Group VB element, such as arsenic (As) and antimony (Sb).

P-type material
A semiconductor material that is formed when intrinsic silicon is mixed with Group IIIB elements, such as indium (In) or gallium (Ga)

Chapter 2: PN Junction Diodes – Chapter Outline

Introduction
Objectives
Key Terms
2.1 P-N Junction Diode Construction
    Self-Examination
2.2 Junction Biasing
    Reverse BiasingForward Biasing
    Self-Examination
2.3 Diode Characteristics
    Forward Characteristics
    Reverse Characteristics
    Combined I–V Characteristics

Chapter 2 Objectives

After studying this chapter, you will be able to:
2.1 Describe the physical characteristics of a P-N junction diode.
2.2 Distinguish between the forward and reverse characteristics of a diode.
2.3 Interpret the I–V characteristics curve of a solid-state diode.
2.4 Interpret a manufacturer's data sheet for a P-N junction diode.
2.5 Evaluate the condition of a diode as good, shorted, or open.

Chapter 2 Key Terms

anode
barrier potential
bias voltage
cathode
depletion zone
diffusion
forward biasing
junction
junction capacitance
knee voltage
leakage current
reverse biasing

switching time
zener breakdown

Chapter 2 – P-N Junction Diodes – Figure List

Figure 2-1. Crystal structure of a junction diode. The crystal structure is continuous, allowing electrons to move readily through the entire structure.

Figure 2-2. Semiconductor materials. Holes are the majority carriers and electrons the minority carriers in P-type material. Electrons are the majority carriers and holes the minority carriers in N-type material.

Figure 2-3. Depletion-zone formation. The depletion zone is created from electrons leaving the N-type material and entering the P-type material to fill holes.

Figure 2-4. Barrier potential of a diode. Barrier potential is a small voltage developed across the P-N junction due to diffusion of holes and electrons.

Figure 2-5. Reverse-biased diode. The negative terminal of the battery is connected to the P-type material, and the positive terminal is connected to the N-type material.

Figure 2-6. Depletion zone creation in a reverse-biased diode. In the N-type material, electrons are pulled toward the positive battery terminal, leaving positive ions in their place. Electrons leave the negative battery terminal and enter the P-type material, filling holes and creating positive ions. The end result is a wider depletion zone.

Figure 2-7. Minority current carriers in a reverse-biased diode. Minority carriers of each material are pushed through the depletion zone to the junction, causing a very small amount of leakage current.

Figure 2-8. Forward-biased diode. The positive battery terminal is connected to the P-type material, and the negative battery terminal is connected to the N-type material.

Figure 2-9. Current carrier flow in a forward-biased diode.

Figure 2-10. A–Forward I–V characteristics of a diode. B–Diode characteristic test circuit. Silicon diode connected in forward conduction. Forward current depends on the source voltage and the value of the current-limiting resistor.

Figure 2-11. A–Reverse I–V characteristics of a diode. B–Reverse diode characteristic test circuit.

Figure 2-12. Forward and reverse I–V characteristics of a diode.

Figure 2-13. Silicon and germanium diode characteristics. Which type of diode requires less forward voltage to go into conduction? Which type of

diode remains in a stable state in the reverse bias until the breakdown voltage is reached.

Figure 2-14. I–V characteristics of a silicon diode at 100°C, 50°C, and 25°C. The germanium diode requires less forward voltage to go into conduction The silicon diode remains in a stable state in reverse bias until the breakdown voltage is reached.

Figure 2-15. Junction capacitance in a reverse-biased P-N junction diode.

Figure 2-16. Diode packages – a) plastic and metal diode outlines ("DO"packages); b) low-power diodes; c) medium-power diodes.

Figure 2-17. A–Diode crystal structure. B–Diode symbol and element names.

Figure 2-18. Diode testing with an ohmmeter. A–Forward-bias test connection. B–Reverse-bias test connection.

Chapter 2 Summary

- The term *P-N junction diode* is used to describe the crystal structure of a two-element electronic device made of N-type and P-type materials.
- The P-N junction formed when these materials are joined responds as a continuous crystal.
- Current carriers diffuse across the junction when a P-N junction diode is formed.
- The diffusion process establishes a barrier potential across the junction.
- For germanium the barrier potential is approximately 0.3 V, and for silicon it is 0.7 V.
- When an external source of energy is applied to the P-N material of a diode, it either adds to or reduces the barrier potential of the junction.
- Forward biasing reduces the barrier potential of a junction and causes the diode to be conductive.
- Forward biasing is achieved by connecting the positive side of the source to the P-type material and the negative side of the source to the N-type material.
- Reverse biasing adds to the barrier potential of a diode and causes it to be nonconductive.
- Reverse biasing occurs when the negative side of the source is connected to the P-type material and the positive side of the source is connected to the N-type material.
- The I–V characteristics of a diode shows how it responds when connected in a circuit that is forward biased or reverse biased.

- In the forward-biased direction, conduction occurs at a few tenths of a volt. This causes a very rapid increase in forward current.
- In the reverse-biased direction, there is very little current with an extremely large value of reverse voltage.
- Reverse current is generally temperature dependent.
- When selecting a diode for an application, one must consider its specifications, which include maximum reverse voltage, reverse breakdown voltage, forward current, surge current, maximum reverse current, power dissipation, and reverse recovery time.
- Diode operation is related to temperature.
- Junction capacitance is a characteristic that changes with bias voltage.
- Switching time is a variable that refers to the time it takes a diode to change from one state to another.
- Diodes can be tested and the leads can be identified with an ohmmeter.
- A diode's leads are identified by matching the polarity of the ohmmeter leads with the P-type and N-type materials of the diode.
- When the positive lead of an ohmmeter is connected to the P-type material and the negative lead to the N-type material, a good diode will show a low resistance reading; reversing the ohmmeter leads will show a high resistance reading.
- A shorted diode will indicate low resistance in both directions.
- An open diode will show infinite resistance in both directions.

Chapter 2 Formulas

(2-1) $I_F = V_S - V_K / R_{limit}$ Forward current of a series circuit with a diode.

Chapter 2 Answers

Examples

2-1. 0.0485 A or 48.5 mA

Self-Examination

2.1

1. diode
2. junction
3. N-type, P-type (any order)
4. electrons

5. holes
6. holes
7. electrons
8. diffusion
9. barrier potential
10. 0.3, 0.7
11. depletion zone

2.2

12. diode
13. biasing
14. reverse
15. forward
16. majority
17. nonconductive
18. increase
19. leakage
20. temperature
21. germanium
22. forward
23. current *or* I_{max}

2.3

24. directly
25. across
26. through
27. forward current *or* I_F
28. germanium, silicon
29. zener
30. avalanche

2.4

31. peak inverse voltage
32. forward current
33. tcmperature
34. capacitor
35. dielectric
36. varicap
37. switching time
38. power

2.5

39. heavy
40. low, infinite
41. shorted
42. open
43. polarity

Chapter 2 Glossary

junction
The point where P-type and N-type semiconductor materials are joined.

diffusion
A process in which electrons from the N-type material of a P-N junction diode move readily across the junction to fill holes in the P-type material.

depletion zone
An area near the P-N junction that is void of current carriers.

barrier potential
The voltage that is developed across a P-N junction due to the diffusion of holes and electrons.

bias voltage
An external source of energy applied to a P-N junction.

reverse biasing
Adding external voltage of the same polarity to the barrier potential to increase in the width of the depletion zone and thus hinder current carriers from entering the depletion zone.

forward biasing
Adding an external voltage of the opposite polarity to the barrier potential to reduce the barrier potential and thus cause current carriers to return to the depletion zone.

leakage current
A small amount of current that flows through the depletion zone when a diode is reverse biased.

knee voltage
The point at which the value of forward voltage is great enough to overcome the barrier potential of the P-N junction.

zener breakdown
A physical process that occurs when electrons are pulled from their covalent bonds in a strong electric field.

junction capacitance
The capacitive effect that occurs from the two independent crystal materials of a diode serving as conductor plates and the depletion zone acting as a dielectric material.

switching time
The time it takes to switch from one state to the other (forward to reverse or vice versa).

anode
The electrode that attracts or gathers in electrons. The P-type material of the crystal serves as the anode in a diode.

cathode
The electrode that gives off or emits electrons. The N-type material of the crystal servers as the cathode in a diode.

Chapter 3: Zener Diodes – Chapter Outline

Introduction
Objectives
Key Terms
3.1 Crystal Structure and Symbols
　　Self-Examination
3.2 Zener Characteristics
　　Self-Examination
3.3 Zener Diode Current Ratings
　　Power Dissipation
　　Zener Impedance
　　Self-Examination
3.4 Analysis and Troubleshooting– Zener Diodes
Summary
Formulas
Review Questions
Problems
Glossary
Zener Diode Data Sheets

Chapter 3 Objectives

After studying this chapter, you will be able to:
3.1 Explain the reverse-bias operation of a zener diode.
3.2 Understand the I–V characteristic curve of a zener diode.
3.3 Analyze the operating characteristics of a zener diode.
3.4 Analyze and test a zener diode.

Chapter 3 Key Terms

avalanche breakdown
derating
power dissipation
power dissipation rating
zener breakdown

Chapter 3 – Zener Diodes – Figure List

Figure 3-1. Zener diode symbol and crystal structure.
Figure 3-2. Zener diodes are similar in appearance to P-N junction diodes.
Figure 3-3. A zener diode is connected in a circuit in the reverse-bias direction.
Figure 3-4. Connecting a zener diode in the reverse bias causes the depletion zone to widen. Unlike a P-N junction diode, the zener diode becomes conductive in this state.
Figure 3-5. I–V characteristics **(A)** and a test circuit **(B)** for evaluating the operation of a zener diode under reverse bias conditions..
Figure 3-6. I_Z-V_Z characteristic of a 1N1775 zener diode.
Data Sheet – 1N4370A-1N4372A

Chapter 3 Summary

• Zener diodes are normally connected in a circuit in the reverse-bias direction.
• When power is applied to a reverse-biased zener diode, the depletion zone of the diode becomes wider.
• Zener voltage (V_Z) is the value of reverse-bias voltage applied to a zener diode.
• The junction of a zener diode becomes conductive at a specific zener voltage (V_Z) value and zener knee current (I_{ZK}).

- The zener voltage (V_Z) of the diode remains fairly constant over a large range of zener current (I_Z).
- Voltage regulation is an important application of the zener diode.
- Zener voltage (V_Z) has a tolerance rating that indicates the value of its accuracy.
- Low-tolerance V_Z diodes are more expensive than high-tolerance devices.
- A low-tolerance zener diode has a sharp knee and a very small change in its zener voltage (V_Z).
- Power dissipation (P_D) refers to the ability of a zener diode to dissipate heat.
- The power dissipation (P_D) rating is used to determine the maximum zener current (I_{Zmax}) that a diode can safely conduct.
- The maximum value of zener current that a zener diode can safely conduct can be determined by the formula $I_{Zmax} = P_D / V_Z$.
- Zener impedance (Z_Z) is related to the slight increase in the value of zener voltage (V_Z) when the zener current (I_Z) of an operating zener diode increases.
- Zener impedance can be determined through the formula $Z_Z = \Delta V_Z / \Delta I_Z$.

Chapter 3 Formulas

(3-1) $P = I \times V$ *Power formula.*
(3-2) $I_{Zmax} = [P_D]/[V_Z]$ *Maximum zener current.*
(3-3) Safe operating range $= I_{Zmax} - I_{Zmin}$ *Safe operating range of a zener diode.*
(3-4) $Z_Z = [\Delta V_Z]/[\Delta I_Z]$ *Zener impedance.*
(3-5) $Z_{ZK} = [\Delta V_{ZK}]/[\Delta I_{ZK}]$ *Zener knee impedance.*

Chapter 3 Glossary

avalanche breakdown
P-N junction breakdown that results when thermally generated holes and electrons gain enough energy from a reverse-biased source to produce new current carriers.

zener breakdown
Current carrier production due to the influence of a strong electric field that causes large numbers of electrons in the depletion region to form covalent bonds across a reverse-biased P-N junction.

power dissipation
The ability of a device to change energy from one form to another, give off heat, or use power.

power dissipation rating
An indication of how much heat a device, such as a zener diode, can give off or dissipate.

derating
Making physical changes to a device, such as reducing lead length or mounting the device on a piece of metal, to permit it to handle more power than its normal power dissipation value.

Chapter 4: Power Supply Circuits – Chapter Outline

Introduction
Objectives
Key Terms
4.1 Transformers
    Self-Examination
4.2 Rectifiers
    Half-wave rectifier
    Full-wave rectifier
    Bridge rectifier
    Self-Examination
4.3 Filters
    C-Input Filters
    Inductance Filters
    Pi Filters
    R-C Filters
    Self-Examination
4.4 Voltage Regulator
    Self-Examination
4.5 Dual Power Supplies
    Self-Examination
4.6 Clipper, Clamper and Voltage Multiplier Circuits
    Clipper Circuits
    Clamper Circuits
    Voltage Multiplier Circuits

Self-Examination
Analysis and Troubleshooting
Summary
Formulas
Glossary

Chapter 4 Objectives

After studying this chapter, you will be able to:
4.1 Calculate the output voltage and current capability of a transformer.
4.2 Analyze half-wave, full-wave, and bridge-rectifier circuits.
4.3 Select the appropriate filter for a given application.
4.4 Analyze a zener regulator circuit under various input and load conditions.
4.5 Diagram the current flow of a dual power supply.
4.6 Describe various types and the operation of clipper, clamper and voltage multiplier circuits.
4.7 Troubleshoot a power supply.

Chapter 4 Key Terms

choke
C-input filter
clamper
clipper
dual power supply
filter
L-input filter
pi filter
pulsating dc
rectification
rectifier
ripple frequency
turns ratio
voltage regulator
clipper
clamper
limiter
voltage multiplier

Chapter 4 – Power Supply Circuits – Figure List

Figure 4-1. Functional block diagram of an electronic power supply. The transformer is the first function of this power supply.

Figure 4-2. A–Power supply transformer. B–Schematic symbol.

Figure 4-3. A rectifier is the second function of an electronic power supply. It changes ac to dc.

Figure 4-4. Graphical representations of a half-wave and full-wave rectifier. A–The half wave rectifier uses one alternation of the ac input. B–The full-wave rectifier uses both alternations of the ac input.

Figure 4-5. Half-wave power supplies. A–Positive output. B–Negative output.

Figure 4-6. Simplification of rectifications: A–The positive alternation causes the diode to be forward biased. B–The negative alternation causes the diode to be reverse biased

Figure 4-7. Half-wave rectifier output. A–Pulsating dc output. B–Composite output.

Figure 4-8. Two-diode full-wave rectifier.

Figure 4-9. Conduction of a full-wave rectifier.

Figure 4-10. Bridge rectifier components.

Figure 4-11. Bridge rectifier conduction.

Figure 4-12. Packaged bridge rectifier assemblies. A–Dual-in-line package. B–Sink-mount package. C–Epoxy package. D–Tab-pack enclosure. E–Thermo-tab package.

Figure 4-13. The filter is the third function of an electronic power supply. It minimizes ripple from the pulsating dc waveform produced by the rectifier.

Figure 4-14. Load current versus output voltage comparisons for L-input and C-input filters.

Figure 4-15. C-input filter action. A–Alternation one. B–Alternation two. C–Waveforms for AC input, diode voltage (V_D) and current (I_D), and load resistor voltage (V_{RL}) and current (I_{RL}).

Figure 4-16. Comparison of C-input filtering with half-wave and full-wave rectifiers. A–Half-wave. B–Full-wave.

Figure 4-17. Inductive filtering. Since inductors tend to oppose any change in current flow, a larger load current can be drawn from an inductive filter without causing a decrease in its output voltage.

Figure 4-18. LC filter. Voltage and current are controlled in this filter, thus creating a purer form of dc.

Figure 4-19. Pi filter. A pi filter has very low ripple content only when used with a light load.

Figure 4-20. RC filter. This filter is similar to a pi filter; however, it uses a resistor in place of an inductor and produces a lower dc output voltage and more ripple.

Figure 4-21. The regulator function of a power supply maintains a stable dc output voltage.

Figure 4-22. Zener diode voltage regulator.

Figure 4-23. A 9-V regulated power supply.

Figure 4-24. Dual power supply conduction.

Figure 4-25. Series and shunt clipper circuits.

Figure 4-26. Input and output waveforms for series clippers.

Figure 4-27. Input and output waveforms for shunt clippers.

Figure 4-28. Biased shunt clipper circuits.

Figure 4-29. Diode clamper circuits.

Figure 4-30. Biased clamper circuits.

Figure 4-31. Zener diode clampers.

Figure 4-32. Voltage doubler circuits. (A) Half-wave doubler. (B) Full-wave doubler.

Figure 4-33. Voltage tripler and voltage quadrupler.

Chapter 4 Summary

- The functions of a typical power supply are transformer, rectifier, filter, and regulator.
- The transformer of a power supply is typically used to step down the input voltage.
- The relationship of turns ratio to voltage is $N_{pri} / N_{sec} = V_{pri} / V_{sec}$.
- The relationship of primary and secondary voltage to current is $V_{pri} / V_{sec} = I_{sec} / I_{pri}$.
- A rectifier changes ac to pulsating dc.
- Three types of rectifiers are half-wave, full-wave, and bridge.
- A half-wave rectifier produces a half-wave output.
- Full-wave and bridge rectifiers produce a full-wave output.
- The composite average value of dc output from a half-wave rectifier is 45% of the ac input voltage, minus the voltage drop of 0.6 V across the silicon diode.
- The composite average value of dc output from a full-wave rectifier is 90% of the ac input voltage, minus the voltage drop of 0.6 V across the silicon diode.

- The composite average value of dc output from a full-wave rectifier is 90% of the ac input voltage, minus the voltage drop of 1.2 V across the two silicon diodes.
- Ripple frequency in a half-wave rectifier is equal to the input frequency.
- Ripple frequency of a full-wave and bridge rectifier is twice the input frequency.
- Filters increase voltage and decrease ripple.
- A C-input filter consists of a capacitor in parallel with the load.
- An L-input filter consists of an inductor in series with the load.
- An LC filter consists of an inductor in series with the load and a capacitor in parallel with the load.
- A pi filter consists of a capacitor in parallel with the load followed by an LC filter.
- An RC filter is similar to a pi filter, except it uses a resistor instead of an inductor.
- A voltage regulator is used to achieve a steady output voltage from a power supply.
- A dual power supply is commonly used with integrated circuits.
- A dual power supply provides both negative and positive voltages with respect to ground.
- Clippers or limiters are diode application circuits used to remove part of an ac input signal.
- Clampers or dc restorer circuits are used to establish a dc voltage reference for an ac signal.
- Voltage multipliers are diode application circuits used to produce a dc output voltage that is a multiple of the peak ac input voltage applied to the circuit.
- Series clipper circuits have one diode that is in series with an ac source and a load resistance.
- Negative series clipper circuits remove the negative alternation of an ac input signal.
- Positive series clipper circuits remove the positive alternation of an ac input signal.
- Shunt clipper circuits have a diode in parallel with a load resistance.
- A series current-limiting resistor is used to prevent the diode of a shunt clipper from shorting the signal source to ground when the diode is forward biased.
- A biased clipper circuit is a shunt clipper that has a dc voltage source used to bias the diode.

- Positive clamper circuits shift an ac input signal above a dc reference voltage.
- A negative clamper shifts an ac input signal below a dc reference voltage.
- A biased clamper may be used to provide an output waveform above or below a dc reference above (or below) a dc reference voltage.
- Zener clampers use zener diode in series with a P-N junction diode to establish a dc reference voltage.
- Negative clampers produce positive dc reference voltage and positive clampers produce a negative dc reference voltage.
- Voltage multipliers produce a dc output voltage that is some multiple of the peak ac input voltage applied to the circuit.
- Half-wave voltage doublers produce a dc output voltage that is approximately twice the peak ac input voltage.
- Full-wave voltage doublers produce an output that is approximately twice the peak ac input voltage applied to the circuit.
- Voltage triplers produce a dc output voltage that is approximately three times the peak ac input voltage applied to the circuit.
- Voltage quadruplers produce a dc output voltage that is approximately four times the peak ac input voltage applied to the circuit.

Chapter 4 Formulas

(4-1) – (4-3) $N_{pri} / N_{sec} = V_{pri} / V_{sec} = I_{sec} / I_{pri}$ Relationship between the turns ratio, voltage ratio, and current ratio.

(4-4) $V_{avg} = 0.637 \times V_P$ The average voltage value of one pulse of dc voltage.

(4-5) composite average value $= 0.318 \times V_P$ Average voltage value of one pulse of dc, taking into account the time of the negative alteration.

(4-6) $V_P = V_{rms} \times 1.414$ The peak value of one ac alternation at the input of a rectifier.

(4-7) composite average value $= (V_{rms} \times 0.45) - 0.6$ V The dc output of an unfiltered half-wave rectifier, less the voltage drop across a silicon diode.

(4-8) composite average value $= (V_{rms} \times 0.9) - 0.6$ V The dc output of a full-wave rectifier, less the voltage drop across a silicon diode.

(4-9) $V_{dc\ (out)} = (V_{rms} \times 0.9) - 1.2$ V The dc output of a full-wave rectifier, less the voltage drop across the silicon diodes.

(4-10) $T = R \times C$ RC time constant.

Chapter 4 Answers

Examples

4-1. 1:5
4-2. 600 V
4-3. 1.2 A
4-4. 6.36 V
4-5. 18 V
4-6. 14.25 V
4-7. 29.1 V
4-8. 28.5 V
4-9. 4 mA

Self-Examination

4.1

1. secondary
2. primary
3. turns ratio
4. voltage, current
5. current, voltage
6. 12
7. 60

4.2

8. one
9. two
10. four
11. positive
12. positive
13. 60
14. 120
15. 120
16. $(10 \text{ V RMS} \times 1.414 \times 0.318) - 0.6\text{V} = 3.89$
    or
    $(10 \text{ V RMS} \times 0.45) - 0.6\text{V} = 3.9$
17. $(30 \text{ V RMS} \times 1.414 \times 0.636) - 1.2 = 25.78$
    or
    $(30 \text{ V RMS} \times 0.9) - 1.2 = 25.8$

4.3

18. voltage, ripple
19. L-input, LC(any order)
20. pi
21. L-input
22. C-input
23. capacitor
24. inductor
25. pi
26. LC

4.4

27. voltage
28. zener diode, resistor
29. reverse
30. zener breakdown
31. series *or* bleeder
32. series *or* bleeder, load

4.5

33. dual *or* split
34. 4, 2, 1, 1
35. positive, negative
36. 3, 2, 1, 2
37. positive, negative

4.6

38. Eliminate one ac alternation at the output
39. Limiter
40. Location of diode relative to R_L
41. Avoid having a short-circuited power source
42. Shunt diode
43. Establish a dc reference voltage
44. dc restorer
45. dc reference voltage
46. ac voltage
47. P-N junction diode
48. Increase dc voltage output
49. two, two

50. two, two
51. three, four
52. four, five

Chapter 4 Glossary

turns ratio
The ratio of the number of turns in the primary winding to the number of turns in the secondary winding.

rectification
The process of changing ac into dc.

rectifier
The function of a power supply that converts ac to dc.

half-wave rectifier
A rectifier in which one alternation of the ac input appears in the output.

pulsating dc
A voltage or current value that rises and falls at a rate or frequency with current flow always in the same direction.

ripple frequency
A circuit that changes pulsating dc into a rather pure form of dc.

C-input filter
A filter circuit employing a capacitor as the first component of its input.

L-input filter
A filter circuit employing an inductor as the first component of its input.

choke
An inductor that is used as a primary filtering element. Refers to the ability of an inductor to reduce ripple voltage and current.

pi filter
A filter with an input capacitor connected to an inductor-capacitor filter, forming the shape of the Greek letter pi.

voltage regulator
A circuit that ensures a stable dc voltage.

dual power supplies
A power supply that has both negative and positive output with respect to ground.

Clipper
A diode circuit that is used to eliminate a portion of a waveform.

Clamper
Also called a dc restorer, a diode circuit used to set or restore the dc reference voltage of a waveform.

Voltage multiplier
A diode circuit used to produce a dc output voltage that is a multiple of the peak ac input voltage.

Chapter 5: Special Semiconductor Diodes – Chapter Outline

Introduction
Objectives
Key Terms
5.1 Tunnel Diodes
    Self-Examination
5.2 Varactor Diodes
5.3 Miscellaneous Diodes
    Schottky-Barrier Diodes
    PIN Diodes
    Gunn Effect Diodes
    IMPATT Diodes
    Self-Examination
Summary
Formulas
Review Questions
Glossary
Special Semiconductor Data Sheets

Chapter 5 Objectives

After studying this chapter, you will be able to:
5.1 Describe the characteristics and applications of tunnel diodes.
5.2 Describe the characteristics and applications of varactor diodes.
5.3 Describe the distinguishing characteristics of Schottky, PIN, IMPATT, and Gunn-effect diodes.

Chapter 5 Key Terms

bistable
Gunn-effect diode
hot carriers
IMPATT diode
negative resistance
PIN diode
quality factor
Schottky-barrier diode
tunnel diode

Chapter 5 – Special Semiconductor Diodes – Figure List

Figure 5-1. Tunneling electrons. The thin depletion zone of a tunnel diode enables current carriers to pass through the barrier with a small amount of bias voltage.
Figure 5-2. I–V characteristics of a tunnel diode.
Figure 5-3. Tunnel diode symbols and crystal structure.
Figure 5-4. Tunnel diode package. 4-4, USD
Figure 5-5. Tunnel diode data. 4-5, USD
Figure 5-6. LC tank circuit and damped oscillatory waveform. 4-6, USD
Figure 5-7. A tunnel diode oscillator. 4-7, USD
Figure 5-8. Bistable switching circuit and operating characteristics. 4-8, USD
Figure 5-9. Diode capacitance values with ambient temperature, T_A, at 25°C. 4-9, USD
Figure 5-10. Varactor diode symbols. 4-11, USD or 10-29, EE5
Figure 5-11. Varactor *LC* circuit. 4-12, USD or 10-30, EE5
Figure 5-12. Schottky-barrier diode construction and symbol. 4-13, USD
Figure 5-13. PIN diode crystal structure and symbols. 4-14, USD
Figure 5-14. PIN diode package types. 4-15, USD

Chapter 5 Summary

- A tunnel diode is a two-element semiconductor with construction similar to that of a conventional silicon diode.
- The P-N materials of a tunnel diode are rather heavily doped, which causes the depletion zone to be very thin in comparison to that of a conventional diode.

- Due to the increased number of current carriers and the thin barrier that exists in a tunnel diode, electrons tend to pass through the barrier with an extremely small amount of energy.
- Applications of the tunnel diode are primarily restricted to high-frequency signal control and switching.
- A varactor diode is manufactured with a very light dose of impurities near the junction and an increased number of impurities away from the junction.
- The depletion zone serves as the dielectric material, and the semiconductors respond as the plates of a capacitor; the value of capacitance can be changed according to the bias voltage.
- The depletion zone of a varactor diode serves as the dielectric material, and the semiconductors respond as the plates of a capacitor.
- The value of capacitance in a varactor diode can be changed according to the bias voltage; thus, this device can respond as a voltage variable capacitor.
- Due to its construction, a varactor diode experiences a higher value of internal capacitance than a conventional silicon diode.
- Applications of the varactor diode are primarily restricted to frequency control.
- Television, FM radio, and automatic frequency control circuits use varactor diodes.
- Schottky diodes are two-terminal devices that are constructed of metal and a piece of semiconductor material.
- The electrons of the N-type material in a Schottky diode possess a rather high level of energy compared with the electrons of the metal.
- When the metal and N-type semiconductor material of a Schottky diode are joined during the forming process, the injected electrons from the N-type material become hot carriers.
- The construction of a Schottky diode causes it to respond entirely to majority current carriers.
- Switching frequencies approaching 20 GHz are common with a Schottky diode.
- A PIN diode is constructed of an intrinsic layer of semiconductor material placed between the P-type and N-type materials of the junction.
- The resistance of a PIN diode can be varied from 10,000 Q to less than 1 Q by control of the current passing through the device.
- PIN diodes have the ability to respond as resistors to high radio frequencies.
- PIN diodes are widely used as high-speed switching devices in microwave control circuits.

- The Gunn-effect diode does not have a distinct P-N junction in its construction; it has a piece of semiconductor material connected between two metal terminals.
- When a particular dc voltage is applied to a Gunn-effect diode, the diode responds by producing a negative resistance.
- The negative resistance produced by a Gunn-effect diode can be used to amplify or generate microwave signals.
- The operating frequencies of a Gunn-effect diode are in the range of 5 GHz to 100 GHz.
- IMPATT diodes are high-frequency devices.
- The term *IMPATT* refers to impact avalanche and transit time.
- IMPATT diodes operate in the reverse-bias region near the avalanche point of conduction.
- A small change in reverse voltage causes an IMPATT diode to have a negative resistance characteristic.
- The negative resistance produced by an IMPATT diode can be used to control RF signals in the 2-GHz to 10-GHz range.
- Although these IMPATT diodes have a rather low-efficiency rating, under normal circumstances this efficiency rating is much better than that of other high-frequency devices.

Chapter 5 Formulas

(5-1) $Q = X_C / R_S$ Quality factor of a varactor diode.
(5-2) $Q = 1 / 2\pi f C R_S$ Quality factor of a varactor diode.
(5-3) $f_c = 1 / 2\pi C R_S$ Cutoff frequency of a varactor diode.

Chapter 5 Answers

Self-Examination

5.1

1. thin
2. tunneling
3. Negative resistance
4. valley voltage or V_V
5. two
6. negative resistance
7. tunnel diode

8. forward
9. negative resistance
10. Voltage peak *or* V_P, valley voltage *or* V_V

5.2

11. dielectric
12. conductor plates
13. capacitance
14. reverse
15. decrease
16. increase
17. temperature
18. Silicon
19. power
20. tuning
21. operating efficiency
22. cutoff
23. high

5.3

24. Schottky
25. electrons
26. switching
27. hot carriers
28. PIN
29. resistor
30. semiconductor
31. avalanche

Chapter 5 Glossary

tunnel diode
A two-element semiconductor device similar in construction to the conventional silicon diode, but which contains a high concentration of impurities.

negative resistance
An electronic condition in which an increase in voltage across a resistive element causes a reduction in current, and vice versa. Certain semiconductors have a negative resistance region in their operation.

bistable
Two stable or stationary operating conditions of an electronic switching circuit.

quality factor
A ratio of the amount of energy stored compared with the actual amount of energy used by the capacitor in the storing process.

Schottky-barrier diode
A semiconductor device in which a barrier is formed between metal and a semiconductor material. This barrier achieves rectification but avoids slowing down the current carriers, so that high-frequency ac can be rectified.

hot carriers
The electrons that are injected into a Schottky-barrier diode.

PIN diode
A semiconductor device that contains a layer of undoped, or intrinsic, semiconductor material between layers of heavily doped P-type and N-type materials.

Gunn-effect diode
A semiconductor device that is used to control high-frequency ac. It is primarily designed to operate in the microwave region.

IMPATT diode
An avalanche diode used as a high-frequency oscillator or amplifier. The negative resistance of this device depends on the transient time of current carriers through the depletion layer.

Chapter 6: Bipolar Junction Transistors—(BJTs) – Chapter Outline

Introduction
Objectives
Key Terms
6.1 BJT Construction
    Self-Examination
6.2 BJT Operation
    NPN Biasing
    PNP Biasing
    Transistor Beta

Chapter 6 Objectives

After studying this chapter, you will be able to:
6.1 Describe the physical construction of NPN and PNP bipolar junction transistors.
6-2 Explain the fundamental operation of a bipolar junction transistor.
6.3 Predict how a bipolar junction transistor will respond in different regions of operation.
6.4 Evaluate the condition of a bipolar junction transistor.
6.5 Analyze and Troubleshoot bipolar junction transistors.

Chapter 6 Terms

active region
base
beta
bipolar
collector
cutoff region
emitter
epitaxial growth
gain

mesa transistor
NPN transistor
planar transistor
PNP transistor
saturation region

Chapter 6 – Bipolar Junction Transistors (BJTs) – Figure List

Figure 6-1. PNP transistor symbol and crystal structure.
Figure 6-2. NPN transistor symbol and crystal structure.
Figure 6-3. PNP alloy-junction transistor formation.
Figure 6-4. Transistor formed by the diffusion process.
Figure 6-5. Typical transistor packages. A−Small-signal. B−Large-signal or power. C−Epoxy. 5-1, USD
Figure 6-6. Emitter-base biasing.
Figure 6-7. Base-collector biasing.
Figure 6-8. NPN transistor biasing.
Figure 6-9. Current carriers passing through an NPN transistor. 12-6, EE5
Figure 6-10. Current carriers of a PNP transistor. 12-7, EE5
Figure 6-11. Collector family of characteristic curves for an NPN transistor. 12-8, EE5
Figure 6-12. Operating regions of a transistor. 12-9, EE5
Figure 6-13. Transistor operation by regions. A−Active region. B−Cutoff region. C−Saturation region. 12-10, EE5
Figure 6-14. Transistor characteristic curve circuit. 12-11, EE5
Figure 6-15. Junction polarity of PNP and NPN transistors. 12-13, EE5
Figure 6-16. Ohmmeter transistor testing. 12-14, EE5
***Figure 6-17.** Flowchart for identifying the base lead of a transistor. NEW / PDF
Figure 6-18. Transistor gain test. A−PNP test circuit. B−NPN test circuit. 12-15, EE5
Figure 6-19. NPN gain test. 12-16, EE5
Figure 6-20. PNP gain test. 12-17, EE5
NPN transistor datasheet—2N3904.
PNP transistor datasheet—2N4403.

Chapter 6 Summary

- Bipolar junction transistors are three-element devices made of semiconductor materials.
- Bipolar junction transistors have two P-N junctions.
- Current flow is controlled in a bipolar junction transistor by altering the voltage applied to the two P-N junctions.
- A bipolar transistor has a thin layer of doped semiconductor material placed between two layers of doped semiconductor material of the opposite polarity.
- The designations NPN or PNP denote the polarity of the semiconductor material used in its construction.
- A bipolar junction transistor is represented by a schematic symbol that has three leads: emitter, base, and collector.
- When the arrow in a bipolar junction transistor symbol points in toward the base (a straight line), it indicates that the transistor is a PNP type.
- When the arrow in a bipolar junction transistor symbol points away from the base, the symbol designates an NPN device.
- The point-contact method of BJT construction has semiconductor materials connected together by pointed wires that are fused to the material.
- Alloy-junction BJTs are formed by attaching two small pieces of metal on opposite sides of a thin piece of semiconductor material.
- The diffusion technique of manufacturing BJTs involves the movement of N-type and P-Type impurity atoms into a piece of silicon.
- For a transistor to function, it must have electrical energy applied to its electrodes; the emitter-base junction is forward biased, and the collector-base junction is reverse biased.
- A forward-biased emitter-base junction causes a large amount of current to flow into the base region.
- The term *alpha* (α) is defined as the ratio I_C / I_E.
- The term *beta* (β) is an expression of current gain and is defined as I_C / I_B.
- Transistor testing, lead identification, and polarity of the material from which it is constructed can be determined with an ohmmeter.
- The two junctions of a transistor are tested as diodes.
- Lead and material polarity identification of a transistor is achieved by using the ohmmeter's voltage source to bias the transistor into operation.
- When identifying the lead and material polarity of a transistor, a base lead is assumed; one ohmmeter lead is attached to this lead and the other lead is switched between the two remaining leads.

Chapter 6 Formulas

(6-1) $I_E = I_B + I_C$ Emitter current.
(6-2) $\alpha = I_C / I_E$
(6-3) $\beta_{dc} = I_C / I_B$ Dc beta.
(6-4) $\beta_{ac} = \Delta I_C / \Delta I_B$ Ac beta.

Chapter 6 Glossary

bipolar
Refers to the conduction by both holes and electrons in bipolar junction transistors.

PNP transistor
A transistor that has a thin layer of N-type material placed between two pieces of P-type material.

emitter
A section of a bipolar transistor that is responsible for the release of majority current carriers.

base
A thin layer of semiconductor material between the emitter and collector of a bipolar transistor.

collector
A section of a bipolar transistor that collects majority current carriers.

NPN transistor
A transistor that has a thin layer of P-type material placed between two pieces of N-type material.

epitaxial growth
A transistor fabrication technique in which atoms are formed on a surface so that they are an extension of the original crystal structure.

planar transistor
When the epitaxial growth process is used to create a transistor and the completed structure has a flat top or level plane surface.

mesa transistor
When the epitaxial process is used to create a transistor and the completed structure rises above the primary surface, forming a plateau.

gain
A ratio of voltage, current or power output to corresponding input values.

beta
A designation of transistor current gain determined by I_C / I_B.

active region
An area of transistor operation between cutoff and saturation.

cutoff region
A condition or region of transistor operation in which current carriers cease or diminish.

saturation region
A condition or region of transistor operation in which a device conducts to its fullest capacity.

Chapter 7: Bipolar Transistor Amplification – Chapter Outline

Introduction
Objectives
Important Terms
7.1 Amplifier Principles
　　Reproduction and Amplification
　　Voltage Amplification
　　Current Amplification
　　Power Amplification
　　Self-Examination
7.2 Basic BJT Amplifiers
　　Self-Examination
7.3 Load Line Analysis
　　Power Dissipation Curve
　　Static Load Line
　　Dynamic Load Line
　　Linear and Non-Linear Operation
　　Classes of Amplification
　　Self-Examination
7.4 Transistor Circuit Configurations
　　Common Emitter Amplifier
　　Common Base Amplifier

Common Collector Amplifier
Self-Examination
7.5 Analysis and Troubleshooting BJT Amplifiers
Amplifier Circuit Troubleshooting
Summary
Formulas
Review Questions
Problems
Terms

Chapter 7 Objectives

After studying this chapter, you will be able to:

7.1 Explain the meaning of amplification with respect to voltage, current, and power.

7.2 Analyze the operation of a basic amplifier.

7-3 Analyze the operation of a basic amplifier by the load-line method.

7.4 Analyze common-emitter, common-base, and common-collector circuit configurations.

7.5 Analyze and troubleshoot an BJT amplifier circuit.

Chapter 7 Key Terms

alpha
bypass capacitor
current amplifier
current gain
emitter biasing
fixed biasing
linear
load line
nonlinear distortion
overdriving
power gain
Q point
self-biasing
self-emitter bias
static state
thermal stability

voltage amplifier
voltage gain

Chapter 7 – Bipolar Junction Transistor Amplification – Figure List

Figure 7-1. Amplification and reproduction. Top−Reproduction. Middle−Amplification. Bottom−Reproduction.
Figure 7-2. Basic bipolar amplifier.
Figure 7-3. Static operating condition of a basic amplifier.
Figure 7-4. Amplifier with ac signal applied.
Figure 7-5. I_B and V_{BE} conditions. A−Static. B−Dynamic. 13-5, EE5
Figure 7-6. Ac amplifier operation. 13-6, EE5
Figure 7-7. Methods of beta-dependent biasing. 13-7, EE5
Figure 7-8. Methods of beta-independent biasing. 13-8, EE5
Figure 7-9. Circuit and characteristic curves of a circuit to be analyzed. A−Circuit. B−Characteristic curves. 13-9, EE5
Figure 7-10. Dynamic load line. A−Circuit. B−Characteristic curves. 13-11, EE5
Figure 7-11. Amplifier operation. 13-12, EE5
Figure 7-12. Overdriven amplifier. 13-13, EE5
Figure 7-13. Classes of amplification. 13-14, EE5
Figure 7-14. Common-emitter amplifier. 13-15, EE5
Figure 7-15. Common-base amplifier. 13-16, EE5
Figure 7-16. Common-collector amplifier. 13-17, EE5

Chapter 7 Summary

- The reproduction function produces an output signal the same size and shape as the input signal.
- The amplification function produces an output signal that is greater in amplitude than the input signal; however, the output signal does not resemble the input signal in shape.
- Voltage amplification is a process where the output signal voltage is made greater than the input signal voltage.
- Current amplification is a process where the output signal current is made greater than the input signal current.
- Power amplification deals with a combination of voltage and current gain in a transistor amplifier.

- In the operation of a bipolar transistor amplifier, the emitter-base junction must be forward biased and the collector-base reverse biased.
- A bipolar transistor amplifier that is properly biased has the necessary dc energy applied to be operational and is considered to be in a static, or dc operating, state.
- Specific operating voltages are selected for an amplifier circuit that will permit amplification.
- If amplification and reproduction are to be achieved, the amplifying device must operate in the center of its active region.
- When an ac signal is applied to the input of a bipolar transistor amplifier, it causes the dc base current and dc collector current to change at an ac rate.
- Capacitors placed in the signal path block dc and pass ac, permitting only the ac signal to be amplified.
- The operation of an amplifier can be analyzed graphically through the use of a load line drawn on a family of collector curves.
- A load line represents two extreme conditions of operation: saturation and cutoff.
- Amplifier operation can be predicted by establishing a Q point and projecting lines to the collector current and collector-emitter voltage values.
- Selection of the operating, or Q, point shows if amplification will be linear or if distortion will occur.
- Bipolar transistor amplifiers are classified according to their bias operating point.
- Three general groups of amplifiers are class A, class B, and class C.
- Amplifier classification is related to the shape of the resulting output waveform.
- The application of a specific amplifier determines the classification used.
- The common-emitter circuit has the input connected to the emitter-base junction and the output taken from the emitter-collector.
- The output voltage developed across the collector of a common-emitter circuit is inverted 180°.
- The common-base circuit has the input connected to the emitter-base junction and the output connected to the base-collector junction.
- The current gain of a common-base circuit is called alpha.
- The input and output of a common-base circuit are in phase.
- The common-collector circuit has the input connected to the base-emitter junction and the output removed from the emitter-collector junction.
- The input and output of a common-collector circuit are in phase.

Chapter 7 Formulas

Gain

(7-1) $A_V = V_{out} / V_{in}$ Voltage gain.
(7-2) $A_V = \Delta V_{out} / \Delta V_{in}$ Change in voltage gain.
(7-3) $A_i = I_{out} / I_{in}$ Current gain.
(7-4) $A_i = \Delta I_{out} / \Delta I_{in}$ Change in current gain.
(7-5) $A_P = P_{out} / P_{in}$ Power gain.
(7-6) $A_P = A_v \times A_i$ Power gain.
(7-7) $I_B = V_{CC} / R_B$ Base current.

Basic amplifier analysis

(7-8) $I_C = \beta \times I_B$ Collector current.
(7-9) $V_{RL} = I_C \times R_L$ Load resistor voltage.
(7-10) $V_{CE} = V_{CC} - V_{RL}$ Collector-emitter voltage.

Power dissipation curve

(7-11) $I_C = P_D / V_{CE}$ Collector current.

Static load line

(7-12) $I_C = V_{CC} / R_L$ Collector current.
(7-13) $I_B = V_{CC} / R_B$ Base current.

Alpha

(7-14) $\alpha = I_C / I_E$ Alpha.

Beta

(7-15) $B_{dc} = I_C / I_B$
(7-16) $B_{ac} = \Delta I_C / \Delta I_B$

Chapter 7 Answers

Examples

7.1. $I_B = 0.1$ mA, $I_C = 9$ mA, $V_{RL} = 4.5$V, $V_{CE} = 4.5$ V
7.2. 2.5 V
7.3. 20 mA, 16.66 mA, 14.29 mA
7.4. 2.1 MΩ, 700 (approximately)
7.5. $\Delta I_C = 3.5$ mA, $\Delta I_B = 20\mu$A, $\beta_{ac} = 180$, $\Delta V_{CE} = 3V$, $A_{V(ac)} = 30$

Self-Examination

7.1

1. input
2. output
3. Reproduction
4. Amplification
5. voltage
6. current
7. power

7.2

8. forward, reverse
9. collector current, *or* I_C
10. 8 mA
11. static
12. dynamic
13. input, output
14. thermal
15. collector
16. collector

7.3

17. Graphically
18. power dissipation, *or* P_D
19. power dissipation, *or* P_D
20. saturation, cutoff (any order)
21. center
22. saturation, cutoff (any order)
23. Linear
24. distorted
25. A
26. A
27. B
28. C

7.4

29. base
30. collector
31. 180°

32. beta
33. emitter, collector
34. in
35. alpha
36. less
37. base, emitter
38. in
39. less

Chapter 7 Glossary

voltage amplifier
A system designed to develop an output voltage that is greater than its input voltage.

voltage gain
A ratio of the output signal voltage to the input signal voltage, which is expressed as $A_v = V_{out} / V_{in}$.

current amplifier
An amplifying system designed to develop output current that is greater than the input current.

current gain
The ratio of output current to input current, which is expressed as $A_i = I_{out} / I_{in}$.

power gain
A ratio of the developed output signal power to the input signal power, which is expressed as $A_p = P_{out} / P_{in}$.

static state
A dc operating condition of an electronic device with operating energy but no signal applied.

thermal stability
The condition of an electronic device that indicates its ability to remain at an operating point without variation due to temperature.

fixed biasing
A type of biasing that is very sensitive to changes in temperature. The resulting output of the circuit is difficult to predict.

self-biasing
A type of biasing in which bias current is used to counteract changes in temperature.

emitter biasing
A type of biasing that improves thermal stability.

bypass capacitor
A capacitor that provides an alternate path around a component or to ground.

self-emitter bias
A type of biasing that is a combination of self-biasing and emitter biasing. The output has reduced gain and the circuit, good thermal stability. However, this type of emitter biasing is not very effective when used independently.

load line
A line drawn on a collector family of characteristic curves that shows how a device will respond in a circuit with a specific value of load resistor.

Q point
An operating point for an electronic device that indicates its dc or static operation with no signal applied.

linear
An amplifying circuit that operates in the active region of a collector family of characteristic curves. It provides signal amplification and duplication.

nonlinear distortion
A characteristic of an output signal in which the amplifying circuit is operating near the saturation or cutoff regions.

overdriving
The effect of inputting a signal that swings into the saturation region during the positive alternation and the cutoff region during the negative alternation. This distorts both alternations of the output.

alpha
Current gain of a common-base amplifier, which is expressed as $\alpha = I_C / I_E$.

Chapter 8: Field Effect Transistors—(FETs) – Chapter Outline

Introduction
Objectives

Key Terms
8.1 Junction Field Effect Transistors
    N-Channel JFETs
    P-Channel JFETs
8.2 JFET Characteristic Curves
    Developing JFET Characteristic Curves
    Load Line analysis
    Dynamic Transfer Curve
    Self-Examination
    Complementary MOS
8.3 Metal Oxide Semiconductor FETs
    E-MOSFETs
    E-MOSFET Operation
    E-MOSFET Characteristic Curves
    D-MOSFETs
    D-MOSFET Operation
    D-MOSFET Characteristic Curves
    V-MOSFETs
    V-MOSFET Operation
    V-MOSFET Characteristic Curves
    MOSFET Handling Procedures
    Self-Examination
    Analysis and Troubleshooting FETs
    Data Sheet Analysis
    FET Testing—JFETs
    MOSFET Testing
    JFET Troubleshooting
Summary
Formulas
Review Questions
Problems
Terms

Chapter 8 Objectives

After studying this chapter, you will be able to:
8.1 Analyze the operation of a junction field-effect transistor.
8.2 Analyze the operation of E-MOSFET, D-MOSFET, and V-MOSFET transistors.
8.3 Analyze and troubleshoot field-effect transistors.

Chapter 8 Key Terms

channel
depletion metal-oxide semiconductor field-effect transistor (D-MOSFET)
drain
dynamic transfer curve
enhancement metal-oxide semiconductor field-effect transistor (E-MOSFET)
gate
junction field-effect transistor (JFET)
ohmic region
pinch-off region
source
substrate
transconductance
vertical metal-oxide semiconductor field-effect transistor (V-MOSFET)

Chapter 8 – Field Effect Transistors (FETs) – Figure List

Figure 8-1. Basic JFET crystal structure and symbol.
Figure 8-2. JFET packages. (Courtesy of Fairchild Semiconductor).
Figure 8-3. N-channel JFET crystal structure, element names, and schematic symbol.
Figure 8-4. P-channel JFET crystal structure, element names, and schematic symbol.
Figure 8-5. A drain family of characteristic curves for a JFET.
Figure 8-6. A drain family of characteristic curves test circuit for an N-channel JFET. 12-22, EE5
Figure 8-7. A drain family of characteristic curves and JFET amplifier circuit. A–N-channel JFET amplifier circuit. B–Drain family of characteristic curves. . 7-6, USD
Figure 8-8. Dynamic transfer curve created from the 10 V V_{DS} data from Figure 8-7. . 7-7, USD
Figure 8-9. E-MOSFET crystal structures, element names, and schematic symbols. A–P-channel E-MOSFET. B–N-channel E-MOSFET. 12-23 A / B, EE5
Figure 8-10. E-MOSFET voltage polarities. A–P-channel. B–N-channel. 7-9, USD
Figure 8-11. A drain family of characteristic curves for an N-channel E-MOSFET. Notice that the pinch-off region and the ohmic region of this device are similar to those of the JFET. 7-10, USD

Figure 8-12. Crystal structures, element names, and schematic symbols of D-MOSFETs. A−N-channel D-MOSFET. B−P-channel D-MOSFET. 7-11, USD

Figure 8-13. A drain family of characteristic curves for an N-channel D-MOSFET. 7-12, USD

Figure 8-14. A cross-sectional view of an N-channel V-MOSFET with schematic symbol. 12-27, EE5

Figure 8-15. A drain family of characteristic curves and a transfer curve for an N-channel V-MOSFET. 7-14, USD

Figure 8-16. Zener diode protection is built into some MOSFETs to avoid electrostatic problems. A−Zener diode gate protection of a MOSFET. B−Transient voltage. 7-26, USD

Figure 8-17. JFET resistance values. . 12-30, EE5

Figure 8-18. D-MOSFET resistance values. . 12-31, EE5

Figure 8-19. E-MOSFET resistance values. . 12-32, EE5

Data sheet for an *N-Channel JFET amplifier*—**BF 245 A**.

Data sheet of a **2N3820** *P-channel JFET*

Data sheet for an *N-channel MOSFET* –**FQN1N60C**

Chapter 8 Summary

- Field-effect transistors (FETs) are unipolar devices.
- The two classifications of field-effect transistors are junction field-effect transistors (JFETs) and metal-oxide semiconductor field-effect transistors (MOSFETs).
- JFETs have three elements: source, drain, and gate.
- The channel of a JFET is a single piece of semiconductor material built on a substrate, and the gate is a piece of semiconductor material diffused into the channel.
- JFET operation is achieved by applying energy to the channel through the source and drain connections.
- A JFET will conduct drain current without gate voltage applied; it is considered a normally on device.
- When the gate on a JFET is reverse biased, it increases the depletion region of the channel and reduces drain current; drain current is therefore controlled by the value of reverse-biased gate voltage.
- JFETs are manufactured as N- or P-channel devices.
- A drain family of characteristic curves is used to analyze the operation of field-effect transistors and show drain current and drain-source voltage for different values of gate-source voltage.

- A load line can be developed that shows how an FET operates in a representative circuit; the extreme conditions of operation are full conduction and cutoff.
- A dynamic transfer curve is developed from drain current and drain-source voltage curves to show input characteristics.
- Transconductance can be determined from a dynamic transfer curve.
- Transconductance is a measure of the ease with which current carriers pass through the channel of a FET.
- The unit of transconductance is the siemens (S).
- MOSFETs are a unique variation of the basic field-effect transistor.
- MOSFETs have the gate insulated from the channel by a thin layer of silicon dioxide; the gate is a strip of metal and not a semiconductor material.
- Operation of an E-MOSFET relies on an induced channel; when the gate is energized, it causes current carriers to move into the channel from the substrate.
- An E-MOSFET is considered to be a normally off device; gate voltage must be applied for it to be conductive.
- D-MOSFETs have a channel formed on the substrate; the gate is a piece of metal insulated from the channel by a thin layer of silicon dioxide.
- D-MOSFET operation is based on energizing the channel and applying voltage to the gate; it is considered to have normally on conduction.
- For an N-channel D-MOSFET, a positive voltage change causes an increase in drain current, and a negative gate-source voltage causes a decrease in drain current.
- A D-MOSFET can be operated as a D- MOSFET or E-MOSFET according to the polarity of the gate voltage.
- V-MOSFETs have a V-groove etched in the surface of the substrate; the channel and gate are then deposited in the groove.
- The construction of a V-MOSFET allows for greater heat dissipation and high-density channel areas.
- The transconductance of V-MOSFETs is quite large compared with that of other FETs; this makes it possible for the V-MOSFET to achieve a great deal more voltage gain.
- MOSFETs are susceptible to damage due to breakdown of the gate insulating material and require special handling.
- MOSFETs are shipped with a shorting strip or placed in conducting foam to prevent electrostatic damage when handling.

• When using MOSFETs, do *not* attach them to energized circuits, solder them into circuits with ungrounded soldering equipment, or handle them without first discharging body static.

Chapter 8 Formulas

(9-1) $I_D = I_{DSS} \left(1 - \frac{V_{GS}}{V_{GS(off)}} \right)^2$

(9-2) Transconductance, gm = $\Delta I_D / \Delta V_{GS}$

Chapter 8 Answers

Examples

8-1. 5mA
8-2. $I_D = 1$ mA, $V_{DS} = 9.3$ V
8-3. $I_D = 1$ mA

Self-Examination

8.1

1. unipolar
2. unipolar
3. channel
4. channel
5. source, drain (any order)
6. source, gate, drain (any order)
7. gate
8. gate, channel (any order)
9. fully
10. on
11. reverse
12. decrease
13. full, cutoff
14. drain current (I_D)
15. negative
16. N
17. reverse
18. operation

19. Transconductance
20. mhos

8.2

21. gate
22. gate
23. gate
24. channel
25. substrate
26. channel
27. induced
28. substrate
29. enhancement
30. electrons
31. depletion
32. holes
33. enhancement
34. gate
35. Depletion
36. Depletion
37. be opposite
38. vertical-groove
39. 0.5
40. enhancement
41. insulation
42. zener

8.3

43. good
44. diode
45. good

Chapter 8 Glossary

junction field-effect transistor (JFET)
A three-element electronic device that bases its operation on the conduction of current carriers through a single piece of semiconductor material, called a channel.

channel
The controlled conduction path of a field-effect transistor.

gate
The control element of a field-effect transistor.

substrate
A piece of underlying N or P semiconductor material on which a device or circuit is constructed.

drain
The output terminal of a field-effect transistor.

source
The region of a field-effect transistor that is similar to the emitter of a bipolar junction transistor.

dynamic transfer curve
A graphic display that shows how a change in input voltage (gate-source voltage) causes a change in output current (drain current).

transconductance
A measure of the ease with which current carriers move through a device.

enhancement metal-oxide semiconductor field-effect transistor (E-MOSFET)
A type of MOSFET with a conduction characteristic in which current carriers are pulled from the substrate into the channel.

pinch-off region
The region of an FET where drain to source voltage is applied to affect the flow of drain current.

ohmic region
The linear constant resistance operating region of an FET.

depletion metal-oxide semiconductor field-effect transistor (D-MOSFET)
A type of MOSFET that has a channel formed on the substrate. The gate is a piece of metal insulated from the channel by a thin layer of silicon dioxide. Operation is based on energizing the channel and applying voltage to the gate.

vertical metal-oxide semiconductor field-effect transistor (V-MOSFET)
A variation of an E-MOSFET with a higher power handling capability.

Chapter 9: FET Amplifiers – Chapter Outline

Introduction
Objectives
Key Terms
9.1 FET Biasing Methods
    Fixed Biasing
    Voltage Divider Biasing
    Self-Biasing
    Self-Examination
9.2 FET Circuit Configurations
    Common Source Amplifiers
    Common Gate Amplifiers
    Common Drain Amplifiers
    Self-Examination
9.3 Troubleshooting FET Amplifiers
Summary
Formulas
Review Questions
Problems
Terms

Chapter 9 Objectives

After studying this chapter, you will be able to:
9.1 Describe fixed bias, voltage-divider bias, and self-bias FET circuits.
9.2 Analyze common-source, common-gate, and common-drain amplifiers.
9.3 Analyze and troubleshoot FET amplifiers.

Chapter 9 Key Terms

fixed bias
voltage-divider bias
self-bias
common-drain amplifier
common-gate amplifier
common-source amplifier

Chapter 9 – FET Amplifiers – Figure List

Figure 9-1. Methods of fixed biasing. A−N-channel JFET. B−N-channel D-MOSFET. C−N-channel E-MOSFET.

Figure 9-2. Fixed biasing of D-MOSFETs. A−N-channel. B−P-channel.

Figure 9-3. Voltage-divider method of biasing for E-MOSFETs. A−N-channel. B−P-channel.

Figure 9-4. Self-biasing of JFETs. A−N-channel. B−P-channel.

Figure 9-5. Common-source JFET amplifier.

Figure 9-6. Common-gate JFET amplifier.

Figure 9-7. Common-drain JFET amplifier.

Chapter 9 Summary

- JFETs and MOSFETs can be used for amplification.
- Biasing of a JFET circuit is very similar to bipolar junction transistor circuits.
- D-MOSFET biasing is similar to JFET circuits.
- E-MOSFET biasing requires a positive value of V_{GS}.
- Voltage divider bias is commonly used for E-MOSFET circuits.
- With voltage-divider biasing, V_{GS} and V_{DD} are the same polarity.
- Self-biasing is also referred to as source biasing.
- FETs can be connected in three different circuit configurations: common-source, common-gate, and common-drain.
- The common-source amplifier is the most widely used.
- With a common-source amplifier, the signal being processed is applied to the gate-source, and the output signal is developed across the source-drain.
- The output signal of a common-source amplifier is inverted 180°.
- The common-gate amplifier has the input signal applied to the source-gate and the output signal developed across the drain-gate.
- Common-drain amplifiers have the input signal applied to the gate, and the output signal removed from the source.
- Common-drain amplifiers are primarily used as impedance-matching circuits; they can match high-impedance devices to a low-impedance output.
- Common-drain amplifiers are also called *source followers.*

Chapter 9 Formulas

(9-1) Voltage-divider biasing, $V_G = V_{DD} \times \frac{R_2}{R_1+R_2}$

(9-2) Voltage gain for a common-source JFET, $A_v = V_{DS}/V_{GS}$

Chapter 9 Answers
Examples

9.1 V_{RL} = 2.5V, V_{DS} = 2.5 V
9.2 V_G = 8.57V, V_{RL}=22V

Self-Examination

9.1

1. gate
2. enhancement
3. resistor
4. negative
5. zero
6. positive
7. same
8. source

9.2

9. source
10. 180°
11. gate
12. source
13. placevoltage
14. 0A, supply V_{DD}

Chapter 9 Glossary

gate biasing of JFETs
Connecting a voltage supply to the gate of a JFET and a resistor for ensuring that the gate-source junction is reverse biased.

self biasing of JFETs
Connecting a resistor to the gate of a JFET for developing an $I_D R_S$ voltage across the resistor when the drain current I_D . A resistor is connected in the gate circuit (with or without a voltage), so that the gate-source junction is reverse biased.

voltage-divider biasing of JFETs
Reverse biasing the gate-source junction of a JFET by using a resistor network in the gate circuit. A resistor is connected between the supply and

the gate; and another between the gate and the ground. These resistors reverse bias the gate-source junction.

common-source amplifier
An amplifier in which the input signal is applied to the gate-source, and the output signal is taken from the drain source.

common-gate amplifier
An amplifier in which the input signal is applied to the source-gate, and the output appears across the drain-gate.

common-drain amplifier
An amplifier configuration in which the input signal is applied to the gate, and the output signal is removed from the source. Also called a *source follower*.

Chapter 10: Amplifying Systems – Chapter Outline

Chapter 10 Objectives

After studying this chapter, you will be able to:
10.1 Analyze the stages of amplification.
10.2 Select the appropriate coupling component for an amplifying system.
10.3 Explain the operation of amplifying system transducers.
10.4 Design basic amplifier circuits.
10.5 Analyze and troubleshoot amplifier systems.

Chapter 10 Key Terms

bel (B)
bode plot
capacitive coupling
cascade
common logarithm
crystal microphone
placeDarlington amplifier
decibel (dB)
direct coupling
dynamic microphone
impedance ratio
intermediate range speaker
logarithmic scale
mantissa
microphone
output transformer
piezoelectric effect
stage of amplification
transducer
transformer coupling
tweeter
voice coil
woofer

Chapter 10 – Amplifying Systems – Figure List

Figure 10-1. Amplifying system.
Figure 10-2. Three-stage voltage amplifier.

Chapter 10 Summary

- The major functions of an amplifying system are energy conversion, amplification, and power distribution.
- Amplifying systems typically have many stages of amplification.
- A stage of amplification consists of an amplifying device and associated components.
- The last stage of amplification in an amplifying system is typically a power amplifier.
- A transducer is a device that changes energy of one form to energy of another form.
- The term *cascade* refers to a series of amplifiers where the output of one stage of amplification is connected to the input of the next stage of amplification.
- The total gain of an amplifying system is the product of the individual amplifier gains.
- Power gain is generally used to describe the operation of the last stage of amplification.
- Sound amplifying systems are usually evaluated on a logarithmic scale because the human ear does not respond to sound levels in the same manner as an amplifying system.
- A logarithmic scale is a non-linear measurement scale where each major division is a whole-number multiple of the previous major division.
- A common logarithm is expressed in powers of 10.

- The Bel (B) is the fundamental unit of sound-level gain.
- The decibel is one-tenth of a Bel.
- Power gain in decibels is determined by the formula $A_p = 10 \log (P_{out} / P_{in})$.
- The range of frequencies over which the gain remains relatively constant is designated as the bandwidth.
- The gain of an amplifier over the range of frequencies specified by its bandwidth is designated as the midband gain.
- The frequencies at which the power gain becomes half are designated as the 3dB-down frequencies.
- Amplifiers generally have two 3dB-down frequencies – a lower frequency designated as the low cutoff frequency, and a high frequency designated as the high cutoff frequency.
- A Bode plot displays the power loss/gain relative to the midband gain as a function of the frequency. By doing so the midband gain is plotted on the 0dB or horizontal axis.
- In a multistage cascade connected amplifier the bandwidth of the overall system is obtained by identifying the range of frequencies for which the gain of each individual amplifier remains at 0 dB. The BW of the overall system is the portion of the combined frequency response curve where the individual amplifier bandwidths completely overlap.
- Three methods of coupling amplifier stages are capacitive coupling, direct coupling, and transformer coupling.
- To ensure amplification over a wide range of frequency, the capacitor to be used as a coupling capacitor must have low capacitive reactance (X_C) at its lowest operating frequency.
- In a direct-coupled amplifier system, the output voltage of one amplifier is the same as the input voltage of the second amplifier.
- Direct-coupled amplifiers are very sensitive to changes in temperature.
- Typically, no more than two stages of amplification are direct-coupled.
- In a transformer-coupled amplifier system, the output of one amplifier stage is connected to the primary winding, and the input of the next amplifier stage is connected to the secondary winding.
- A transformer that is used to couple an amplifier to a load device is called an output transformer.
- An input transducer is used to change sound energy into electrical energy.
- An output transducer is used to change electrical energy into sound energy.
- Small high-frequency speakers are called tweeters.
- The cone of a tweeter is generally made of a rather stiff material.

• Large low-frequency speakers are called woofers.
• The cone of a woofer is usually flexible.

Chapter 10 Formulas

(10-1) $A_p = \log (P_{out} / P_{in})$ (in Bels) Power gain.
(10-2) $A_{p(dB)} = 10 \log_{10}(P_{out} / P_{in})$ Power gain in decibels.
(10-3) $P_{out} = P_{in} \times 10^{0.1 \times A_p}$
(10-4) $A_v = 20 \log_{10}(V_{out} / V_{in})$ Voltage gain in decibels.
(10-5) $\text{BW} = f_{3dB-down\ (High)} - f_{3dB-down\ (Low)}$
(10-6) $n = \frac{N_{pri}}{N_{sec}}$
(10-7) $n^2 = \frac{Z_{pri}}{Z_{sec}}$

Chapter 10 Answers

Examples

10.1. 120
10.2. 2.78
10.3. 40
10.4. 42.958 dB
10.5. 19980 Hz or 19.98 kHz
10.6. 2400 Ω

Self-Examination

10.1

1. stage
2. power
3. transducer
4. cascade
5. 60
6. power gain, *or* A_P
7. 10.97
8. high and the low cutoff, or between the 3dB-down(high) and 3dB-down (low) frequency
9. octave
10. loss

10.2

11. capacitive, direct, transformer (any order)
12. large
13. small
14. placeDarlington amplifier
15. placeDarlington
16. load

10.3

17. microphones, antennas, (any order, other transducer possible)
18. crystal, dynamic (any order)
19. dynamic
20. speaker
21. voice coil
22. tweeter, intermediate range (*or* midrange), woofer (any order)

Chapter 10 Glossary

Stage of amplification
A transistor or IC amplifier and all the components needed to achieve amplification.

transducer
A device that changes energy from one form to another. A transducer can be on the input or output.

cascade
A method of amplifier connection in which the output of one stage of amplification is connected to the input of the next stage of amplification.

logarithmic scale
A nonlinear scale of measurement where each major division is a whole-number multiple of the previous major division.

common logarithm
A logarithm that is expressed in powers of 10.

mantissa
Decimal part of a logarithm.

Bel (B)
A measurement unit of gain that is equivalent to a 10:1 ratio of power levels.

decibel (dB)
One-tenth of a Bel. Used to express gain or loss.

Midband gain
The gain of an amplifier where it remains relatively constant over a certain range of frequencies.

Cutoff or 3dB down frequency
The frequency where the power is reduced to half of the midband gain. Usually amplifiers have two cutoff frequencies – one is the low cutoff frequency, and the other is the high cutoff frequency.

bandwidth
The range of frequencies between the high and the low cutoff frequencies.

Bode plot
A plot of the overall gain of an amplifier compared to the midband gain for a certain range of frequencies.

capacitive coupling
A method of coupling amplifier stages with capacitors.

direct coupling
A method of coupling amplifier stages in which the output of one amplifier stage is connected directly to the input of the next amplifier stage.

placeDarlington amplifier
Two transistor amplifiers cascaded together for increasing the overall values of the current gain and input impedance.

transformer coupling
A method of coupling amplifier stages in which the output of one amplifier is connected to the input of the next amplifier by mutual inductance.

output transformer
A transformer that is used to couple an amplifier to a load device.

impedance ratio
The ac resistance ratio between the input and output of a transformer.

microphone
A device that changes sound energy into electrical energy.

crystal microphone
Input transducer that changes sound into electrical energy by the piezoelectric effect.

piezoelectric effect
The property of a crystal material that produces voltage by changes in shape or pressure.

dynamic microphone
An input transducer that changes sound into electrical energy by moving a coil through a magnetic field.

voice coil
The electromagnetic part of a speaker.

tweeter
A speaker designed to reproduce only high frequencies.

woofer
A speaker designed to reproduce low audio frequencies with high quality.

intermediate range speaker
A speaker that is designed to respond efficiently to frequencies in the center of the audio range.

Chapter 11: Power Amplifiers – Chapter Outline

Introduction
Objectives
Key Terms
11.1 Single-Ended Power Amplifiers
    Self-Examination
11.2 Push Pull Amplifiers
    Class B Push Pull Amplifier
    Class AB Push Pull Amplifiers
    Self-Examination
11.3 Complimentary Symmetry Amplifiers
    Self-Examination
11.4 Analysis and Troubleshooting – Power Amplifiers
Summary
Formulas
Review Questions
Problems
Terms

Chapter 11 Objectives

After studying this chapter, you will be able to:

11.1 Analyze the operational efficiency of a single-ended power amplifier.

11.2 Analyze the operational efficiency of class B and class AB push-pull power amplifiers.

11.3 Analyze the operational efficiency of a complementary-symmetry power amplifier.

11 .4 Analyze and troubleshoot power amplifiers.

Chapter 11 Key Terms

complementary-symmetry amplifier
crossover distortion
efficiency
single-ended amplifier
transformer-coupled push-pull transistor amplifier

Chapter 11 – Power Amplifiers – Figure List

Figure 11-1. Single-ended audio amplifier.

Figure 11-2. Collector family of characteristic curves for the single-ended audio amplifier in Figure 11-1.

Figure 11-3. Transformer-coupled class B push-pull amplifier.

Figure 11-4. Combined collector family of characteristic curves for the push-pull amplifier in Figure 11-3.

Figure 11-5. Input characteristic of a silicon transistor.

Figure 11-6. Transistor characteristics and operating point. A – Input characteristic for a silicon transistor. B – Bias operating points on a load line.

Figure 11-7. Transformer-coupled class AB push-pull amplifier.

Figure 11-8. Class B Complementary-symmetry amplifier.

Chapter 11 Summary

- A single-ended amplifier operates as a class A amplifier and has an operating efficiency of approximately 30%.
- Efficiency refers to a ratio of developed output power to dc supply power and is expressed by the formula $\eta = (P_{out} / P_{in}) \times 100\%$.

- A single-ended audio amplifier responds to frequencies from 15 Hz to 15 kHz; this frequency range is known as audio frequency.
- The term *audio* refers to the range of human hearing.
- A push-pull amplifier can operates as a class B or class AB amplifier and has an operating efficiency of about 80%.
- The bias operating point of the two transistors used in a class B push-pull amplifier is at cutoff; therefore, each transistor handles only one alternation of the sine-wave input signal.
- A class AB push-pull amplifier is biased slightly above cutoff; this helps to reduce distortion.
- Complementary-symmetry amplifiers use a PNP and a NPN transistor with similar characteristics.
- Complementary-symmetry amplifiers have an operating efficiency and linearity that is equal to that of a conventional push-pull circuit.
- Component cost and weight are reduced with complementary-symmetry amplifiers because they do not incorporate a transformer.

Chapter 11 Formulas

(11-1) $\eta = (P_{out} / P_{in}) \times 100\%$ Efficiency.

(11-2) $R_1 = R_3 = \beta \times R_L$ R_1 and R_3 values (Complementary-symmetry amplifier).

(11-3) $I_{n\ (in)} = V_{CC} / (R_1 + R_3)$ Network current (complementary-symmetry amplifier).

Chapter 11 Answers

Examples

11.1 2W

11.2 $I_{C(rms)} = 0.141$ A, $P_{in} = 1.697$ W , $\eta = 94.29\%$

11.3 $R_1 = R_3 = 1000\ \Omega$, $R_2 = 200\ \Omega$

Self-Examination

11.1

1. False, the efficiency is low for class A amplifiers
2. single ended amplifier
3. (a) in the middle of the load line

4. (d) impedance matching with the load
5. (a) 0V when the output is shorted

11.2

6. (a) greater efficiency and hence greater output
7. (d) class B has increased crossover distortion
8. (c) distortion owing to change in the biasing
9. (a) Class A amplifiers have the lowest efficiency
10. (b) Class B amplifiers have the highest operating efficiency

11.3

11. emitter follower
12. complementary-symmetry amplifier
13. amplification of the input ac signal is identical for both halves of the waveform.
14. (a) 1 NPN and 1 PNP transistor
15. (b) Non-symmetrical amplification would occur if the transistor characteristics are not matched exactly.

Chapter 11 Glossary

single-ended amplifier
A power amplifier that has only one active device, operates as a class A amplifier, and has an operating efficiency of about 30%.

efficiency
A ratio of developed output power to dc supply power.

transformer-coupled push-pull transistor amplifier
A power amplifier that is somewhat like two single-ended amplifiers placed back to back. The input and output transformers both have a common connection. It operates as a class B amplifier and has an operational efficiency of approximately 80%.

crossover distortion
The distortion of an output signal at the point where one transistor stops conducting and another conducts.

complementary-symmetry amplifier
A power amplifier that uses NPN and PNP transistors having similar operating characteristics, for amplifying each half (alternation) of the ac input waveform. It operates as a class B amplifier and has an operational efficiency of approximately 80%.

Chapter 12: Thyristors – Chapter Outline

Introduction
Objectives
Key Terms
12.1 Silicon Controlled Rectifiers
    SCR Construction
    SCR Operation
    SCR I–V characteristics
    SCR Power Control
    DC Power Control
    AC Power Control
    Self-Examination
12.2 Triacs
    Triac Construction
    Triac Operation
    Triac I–V Characteristics
    Triac DC Power Control
    Triac AC Power Control
    Static Switch
    Self-Examination
12.3 Diacs
    Diac Construction and Operation
    Diac I–V characteristics
    Diac-Triac Power Control
    Self-Examination
12.4 Unijunction Transistors
    UJT Construction
    UJT Operation
    UJT I–V characteristics
    SCR Pulse Triggering with a UJT
    UJT Relaxation Oscillator
    UJT Pulse Trigger Control Circuit
    Self-Examination
12.5 Programmable Unijunction Transistors
    PUT Construction
    PUT Operation
    PUT I–V characteristics
    PUT Pulse Triggering

Chapter 12 Objectives

After studying this chapter, you will be able to:
12.1 Analyze the operation of an SCR in dc and ac power control circuits.
12.2 Analyze the operation of a triac in dc and ac power control circuits.
12.3 Analyze the operation of a diac in ac power control circuits.
12.4 Analyze the operation of a UJT in pulse-triggered circuits.
12.5 Analyze the operation of a PUT in pulse-triggered circuits.
12.6 Analyze the operation of SCS and GTO thyristors.
12.7 Analyze and Troubleshoot thyristors.

Chapter 12 Key Terms

alternation
bidirectional triggering
conduction time
contact bounce
forward breakover voltage
holding current
interbase resistance
internal resistance
intrinsic standoff ratio
latching
negative resistance region
off-state resistance

triggering
turn-on time

Chapter 12 – Thyristors – Figure List

Figure 12-1. SCR crystal. A−Symbol. B−Structure.
Figure 12-2. Representative SCR packages.
Figure 12-3. Equivalent SCR.
Figure 12-4. SCR response.
Figure 12-5. I–V characteristics of an SCR.
Figure 12-6. Gate-current characteristics of an SCR.
Figure 12-7. Dc power control switch. 18-10, EE5
Figure 12-8. Dc power control circuit. 18-11, EE5
Figure 12-9. SCR ac power control switch. 18-12, EE5
Figure 12-10. Phase-control SCR circuit and waveforms: (a) applied ac; (b) 0° delay; (c) 45° delay; (d) 90° delay; (e) 135° delay; (f) 180° delay. 18-13, EE5
Figure 12-11. Junction diagram and schematic symbol of a triac. 18-14, EE5
Figure 12-12. I–V characteristics of a triac. 18-16, EE5
Figure 12-13. Trigger modes of a triac. 8-13, USD
Figure 12-14. Dc control circuit for a triac. A−Base circuit. B−Source voltage and conduction current waveforms. 8-14, USD
Figure 12-15. Ac phase control of a triac. A−With single-leg phase shifter. B−With bridge phase shifter. 8-15, USD
Figure 12-16. Ac control conditions of a triac. A−0° delay. B−45° delay. C−90° delay. D−135° delay. E−165° delay. F−180° delay. 18-20 A-E, EE5
Figure 12-17. Triac static switch circuits. A−Static switch. B−Three-position static switch.
    18-17 A - B, EE5
Figure 12-18. CityplaceCrystal structure and schematic symbol of a diac. 18-23 A-B, EE5
Figure 12-19. I–V characteristics of a diac. 18-23 C, EE5
Figure 12-20. Full-wave triac power control circuit. 8-20, USD
Figure 12-21. Unijunction transistor crystal structure and schematic symbol. 8-22, USD
Figure 12-22. Equivalent of a UJT. 8-23, USD
Figure 12-23. Characteristic curve of a UJT. 8-24, USD
Figure 12-24. Waveforms of a pulse-triggered SCR. 8-21, USD
Figure 12-25. UJT sawtooth oscillator. 8-25, USD

Chapter 12 Summary

- A thyristor is a general classification for solid-state devices that are used for power control applications.
- The term *thyristor* is a contraction of the words (thyr)atron transistor.
- Thyristors can be two-, three-, or four-element devices.
- Two major classifications of thyristors are reverse blocking and bidirectional.
- Thyristors classified as reverse blocking are silicon controlled rectifiers, unijunction transistors, programmable unijunction transistors, gate turn-off switches, and silicon-controlled switches.
- Thyristors classified as bidirectional are triacs and diacs.
- An silicon controlled rectifiers (SCR) is a solid-state device made of four alternate layers of P-type and N-type silicon.
- The leads of an SCR are the anode, cathode, and gate.
- The gate of an SCR is used to trigger conduction current between the anode and cathode.
- When conduction in the SCR reaches the holding current level, the SCR latches and holds its conduction.
- To stop conduction in the SCR, the anode-cathode current must be lowered below the holding current level or the voltage must be momentarily removed.

- When an SCR is used as a power control device, it primarily responds as a switch.
- Dc power control using an SCR requires two switches to achieve control: one turns on the source voltage, and the second switch controls the gate.
- Ac power control using an SCR has automatic commutation or turn-off because of the reversal of each alternation.
- The conduction time of an alternation can be changed to make a circuit have variable ac power control; this is achieved with a phase shifter, an RC circuit, or a pulse-control device.
- A triac is functionally classified as a gate-controlled ac switch; it responds as two reverse-parallel-connected SCRs with one common gate.
- Each terminal connection of the triac is jointly connected to an N-P material combination; the leads of a triac are called terminal 1, terminal 2, and gate.
- Conduction in a triac is achieved by selecting an appropriate crystal combination; selection depends on the polarity of the source.
- The dual polarity of each connection causes the triac to have four possible trigger combinations; the most sensitive modes of operation occur when the gate polarity matches the polarity of terminal 2.
- A triac has bidirectional conductivity; therefore, conduction in quadrants I and III of its I–V characteristics chart is identical.
- When a triac has ac control of the gate, it is possible to achieve 180° delay of the conduction time for each alternation; conduction time of the ac waveform can be varied to achieve 100% control of a power source.
- A diac is a bidirectional diode used to trigger a triac.
- The diac is the equivalent of an NPN transistor with no base connection.
- Conduction for the diac is the same in each direction.
- The I–V characteristic of a diac shows that conduction occurs when the breakover voltage is exceeded; this is the same in quadrants I and III.
- Pulse triggering of an SCR can be accomplished with a unijunction transistor.
- A unijunction transistor (UJT) is commonly described as a voltage-controlled diode; it responds as a three-terminal, single-junction solid-state device.
- A UJT is constructed with a small bar of N-type silicon, which has base 1 and base 2 attached to the ends of the bar; an emitter is formed by fusing an aluminum wire to the approximate center of the bar.
- Operation of a UJT is based on the bias voltage applied to the emitter-base 1 junction; when E-B_l is forward biased, it lowers the internal B_lB_2 resistance.

- The bias voltage needed to achieve conduction in a UJT is determined by the intrinsic standoff ratio.
- UJTs are normally used in pulse-generator circuits that are attached to an SCR; the pulse rate of the circuit is an RC function.
- A programmable unijunction transistor (PUT) is commonly used to generate trigger pulses; it is actually a thyristor that responds as a UJT.
- The intrinsic standoff ratio voltage of a PUT can be altered or programmed externally by changing the resistance ratio of a voltage-divider network.
- The crystal structure of a PUT has its gate connected to the N-type material nearest to the anode; a P-N junction is formed by the anode-gate.
- Conduction in a PUT is controlled by the bias voltage of anode-gate.
- Gate turn-off thyristors are unique power control devices that can be triggered into conduction or turned off by a gate signal.
- The silicon-controlled switch (SCS) is a four-layer PNPN thyristor; an external lead connection is made to all four layers of the device.
- A SCS has an anode, cathode, anode-gate, and a cathode-gate.
- The I–V characteristics for a silicon-controlled switch are the same as those of an SCR; conductivity occurs in quadrant I and occurs when the anode is positive and the cathode is negative.
- Triggering with a SCS is achieved by making the anode-gate negative or the cathode-gate positive.
- An SCS can be turned off in three ways: bring the anode current below the holding current level; apply a negative pulse to the cathode-gate; apply a positive-going pulse to the anode-gate.
- A Gate turn-off thyristor (GTO) is very similar in construction to the SCR—a positive gate pulse causes the device to be conductive, and a negative gate pulse signal turns off the conduction.

Chapter 12 Formulas

(12-1) $\eta = R_{B1} / (R_{B1} + R_{B2})$ Intrinsic standoff ratio.
(12-2) $V_G = \eta \times V_S$ Gate voltage (V_G) of a PUT.
(12-3) $t = (R_2 + R_3) \times C_1$ Discharge time of a sawtooth generator.

Chapter 12 Answers
Examples

12.1. (1) V_{BO} =48V for I_G = 6 mA; (2) V_{BO} =148V for I_G = 4 mA; (3) V_{BO} =325V for I_G = 1 mA; (4) V_{BO} =400V for I_G = 0 mA

12.2. Assuming that the lamp load has already been switched on. To switch it off:

1) Momentarily press the Stop PB
2) Gate of SCR-2 will be triggered
3) V_{BO} of SCR-2 is reduced
4) SCR-2 goes into conduction
5) Capacitor C_1 discharges through SCR-2
6) The anode of SCR-1 is momentarily grounded
7) SCR-1 is switched off
8) The lamp load is switched off.

12.3. No power is developed by the load as the SCR never goes into conduction.

12.4. 1) 7 – 15 mA 2) 5 – 10 mA

12.5. Refer to Figure 12-16c, to see that the triac begins conducting at 90° in the positive alteration of the ac input. This will cause the output to be available from 90° through 180° of this alternation. After this the triac will be switched off, until 270° of the negative alteration of the ac input. The triac begins conducting at 270° and will produce to 360°. This process will then be repeatead.

12.6. In switch position 2, the diode D_1 will be connected to the gate during the positive alteration of the ac input, and this triggers the triac into conduction, energizing the load. In the negative alteration of the ac input, the diode is reverse biased and does not trigger the triac into conduction. The load is thus de-energized during the negative alteration of the ac input.

12.7. In quadrant III, when T_1 is negative and T_2 is positive, the diac does not conduct until V_{BO2}, is reached. After the V_{BO2} value is reached the diac begins to conduct.

12.8. Full power will be developed by the load for this condition of operation. The triac is triggered at the earliest possible instant (at 0°) in the positive half alteration causing it to conduct for the entire alteration. It is triggered again at the earliest possible instant (at 180°) of the negative alteration causing conduction.

12.9. When the voltage applied across the Emitter-Base1 junction reaches V_P, it is forward biased. At this point there is an increase in the current I_E. V_P drops to V_V (the Emitter-Base1 voltage at the 'valley voltage point') even as the current is increases. This portion of the UJT characteristic curve represents the negative resistance region.

12.10. When the voltage applied across the anode-cathode junction reaches V_P, it is forward biased. At this point there is an increase in the current I_{AK}. V_P drops to V_V (the anode-cathode voltage at the 'valley voltage point') even as the current is increases. This portion of the PUT characteristic curve represents the negative resistance region.

12.11. When the PUT goes into conduction it causes the capacitor C_1 to discharge through R_4 causing a momentary pulse to be generated across R_4. The voltage across the capacitor reduces and this causes the anode voltage of the PUT to reduce as well. The PUT turns off which in turn causes C_1 to charge through the resistor R_3.

12.13. t=2.5ms; T 12.5ms

Self-Examination

12.1

1. anode, cathode, gate (in any order)
2. forward
3. three, four
4. low
5. high
6. anode-cathode voltage
7. alternation
8. I
9. forward breakover voltage, *or* V_{BO}
10. III
11. increase
12. forward
13. gate
14. gate
15. holding

12.2

16. SCRs
17. polarity
18. $N_lP_lN_2P_2$
19. $N_3P_2N_2P_l$
20. zero
21. lowers
22. +

23. –
24. gate
25. source, *or* supply

12.3

26. diac
27. T_1, T_2 (any order)
28. I, III (any order)
29. gate
30. R_l, C_1 (any order)
31. conduction, *or* current

12.4

32. resistor
33. three
34. trigger
35. base 1, base 2 (any order)
36. emitter
37. E-B_1
38. decreases
39. negative resistance
40. intrinsic standoff
41. triggering

12.5

42. programmable unijunction transistor, *or* PUT
43. silicon controlled rectifier, *or* SCR
44. gate
45. gate, anode (any order)
46. positive
47. low
48. high
49. intrinsic standoff ratio
50. programmed, *or* changed
51. trigger

12.6

52. silicon controlled rectifier, *or* SCR
53. two
54. +

55. −
56. three
57. +, −
58. gate

Chapter 12 Glossary

internal resistance
The resistance between the anode and cathode of an SCR or between terminals 1 and 2 of a triac. Also called the dynamic resistance.

off-state resistance
The resistance of an SCR or triac when it is not conducting.

triggering
The process of causing an NPNP device to switch states from off to on.

latching
The process of placing an SCR or triac in a holding state in which the device turns on and stays in conduction when the gate current is removed.

alternation
Half of an ac sine wave. There is a positive alternation and a negative alternation for each ac cycle.

turn-on time
The time of an ac waveform when an alternation occurs.

forward breakover voltage
The voltage at which an SCR or triac goes into conduction in quadrant I of its I–V characteristics.

holding current
A current level that must be achieved when an SCR or triac latches or holds in the conductive state.

conduction time
The time when a solid-state device is turned on or in its conductive state.

bidirectional triggering
The triggering or conduction that can be achieved in either direction of an applied ac wave. Diacs and triacs are of this classification.

contact bounce
When a mechanical switch is turned on, the contacts are forced together. This causes the contacts to open and close several times before making firm contact.

interbase resistance
The resistance between base 1 and base 2 of a unijunction transistor.

negative resistance region
A UJT characteristic where an increase in emitter current causes a decrease in emitter voltage.

intrinsic standoff ratio (η)
The ratio of the resistance of Base 1 (R_{B1}) to the total resistance of the bases ($R_{B1} + R_{B2}$) of a UJT.

Chapter 13: Optoelectronic Devices

Introduction
Objectives
Key Terms
13.1 The Nature of Light
    Electromagnetic Spectrum
    Wave Theory
    Quantum Theory
    Terms and Units of Measurement
    Radiometric Systems
    Photometric Systems
    Self-Examination
13.2 Radiation Sources
    Photometric Source Classifications
    Incandescent Lamps
    Light Emitting Diodes
    LED Characteristics
    LED Applications
    Self-Examination
13.3 Optoelectronic Detectors
    Photo-emissive Devices
    Photoconductive Devices
    Photodiodes

PIN Diodes
Phototransistors
Light-Activated SCR
Photovoltaic Devices
Self-Examination
13.4 Analysis and Troubleshooting–Optoelectronic Devices
Summary
Formulas
Review Questions
Problems
Terms

Chapter 13 Objectives

After studying this chapter, you will be able to:
13.1 Describe some of the basic characteristics of light energy.
13.2 Evaluate the performance of a device that radiates light energy.
13.3 Evaluate the performance of a device that detects light energy.
13.4 Analyze and troubleshoot optoelectronic systems.

Chapter 13 Key Terms

angstrom
candela (cd)
dark current
dynodes
electromagnetic spectrum
heterochromatic source
illuminance
infrared emitting diode (IRED)
intensity
light emitting diode (LED)
lumen
luminous exitance
luminous flux
luminous intensity
monochromatic source
panchromatic source

photoconductive device
photoelectric emission
photoemissive device
photometric system
photoresistive cell
photovoltaic cell
photovoltaic device
quantum theory
radiance
radiant exitance
radiant flux
radiant incidence
radiant intensity
radiometric system
steradian (sr)
transverse wave
wavelength

Chapter 13 – Optoelectronic Devices – Figure List

Figure 13-1. Electromagnetic spectrum.
Figure 13-2. Wavelength.
Figure 13-3. Solid angle of a sphere.
Figure 13-4. Spectral response of radiation sources.
Figure 13-5. Light-emitting diode. A−LED symbol. B−LED crystal structure. C−LED package.
Figure 13-6. LED spectral response curves.
Figure 13-7. I–V graph of a GaAs LED.
Figure 13-8. Light output versus forward current characteristic of a GaAsP LED.
Figure 13-9. Current-limiting resistor calculation for an LED.
Figure 13-10. LED display devices. A−Seven-segment LED display with four diodes per segment.
B−Diagram showing connections of segments. C−LED 5 × 7 dot-matrix display. D−LED dot-matrix diagram. **10-10, USD**
Figure 13-11. Optocouplers. A−Package. B−Equivalent circuit. C−Other configurations. **10-11, USD**
Figure 13-12. Photoemissive cell. **10-12, USD**
Figure 13-13. Spectral response curve for a phototube. **10-13, USD**

Chapter 13 Summary

• Optoelectronic devices have electrical properties that are affected by light.

- The term *light* includes visible, infrared, and ultraviolet regions of the frequency spectrum.
- Light is a form of radiant energy.
- Radiant energy is made up of the entire optical spectrum, which includes ultraviolet, visible, and infrared light.
- One theory of radiant energy transmission considers light to travel as transverse waves that cause particles of a medium to vibrate at right angles to the direction of motion of the waves.
- Light travels at 186,000 miles per second, or 300,000,000 meters per second.
- There is an inverse relationship between the length of a wave and its frequency.
- Wavelength is a measure of how far the wave travels during one cycle of operation.
- The quantum theory of light considers light to be emitted in discrete packets of energy called photons.
- The energy of a photon depends on the wavelength of the light.
- Two systems of measurement used by the optoelectronic field are radiometric and photometric.
- The radiometric system deals with the entire optical spectrum.
- The photometric system deals with electromagnetic energy that falls in the visible part of the electromagnetic spectrum.
- Photometric terms are preceded by the word luminous, and radiometric terms are preceded by the word radiant.
- The unit of luminous intensity is the candela.
- An optoelectronic system is distinguished from other systems by its source, which must generate some form of radiant energy, and its detector, which responds to or receives radiant energy.
- Most optoelectronic systems respond to radiant energy that falls in the visible part of the spectrum.
- The source of an optoelectronic system is usually classified according to the amount of visible light emitted; sources can be panchromatic, heterochromatic, or monochromatic.
- An incandescent lamp is an example of a panchromatic source, which generates light over a large part of the visible spectrum.
- A mercury arc lamp is an example of a heterochromatic source; this lamp radiates energy in a narrow part of the spectrum that is red to orange and peaks at 6500 Å.

- A sodium vapor lamp is an example of a monochromatic source; it generates light that is predominantly of one wavelength.
- Light-emitting diodes (LEDs) are solid-state sources that have a P-N junction that emits light when forward biased.
- Material used to make the P-N junction of an LED determines the wavelength of energy released.
- An LED must be connected in series with a current-limiting resistor to prevent excessive current from damaging the junction.
- The detector of an optoelectronic system is designed to change radiant energy into electrical energy.
- The three general categories of detectors are photoemissive, photoconductive, or photovoltaic.
- Photoemissive detectors give off or emit electrons when the light-sensitive material of the cathode absorbs photons of light energy.
- Photomultiplier tubes are photoemissive devices that respond to the secondary emission of several dynode plates, which increase the output of the device.
- Photoconductive detectors change light intensity into electrical conductivity.
- Cadmium sulfide (CdS) cells change resistance when exposed to light.
- Photodiodes are photoconductive devices that have a light-sensitive P-N junction.
- A photodiode is normally connected in the reverse-bias direction; when light is applied to the junction, current carriers return to the depletion region and cause conduction.
- A PIN photodiode has a layer of intrinsic semiconductor material between the P-type and N-type materials; this construction lowers the junction capacitance, which causes it to have a faster response time.
- Phototransistors are photoconductive devices that have two P-N junctions; the collector-base junction responds as a photodiode.
- A phototransistor has an amplification capability, which makes it more sensitive to changes in light intensity.
- Light-activated Silicon Controlled Rectifiers (LASCRs) can be triggered into conduction when light energy is applied to the gate, and voltage of proper polarity is applied to the anode and cathode terminals. Once triggered it will stay latched until the current drops below the holding level.
- Photovoltaic devices are designed to change light energy directly into electrical energy.

• Selenium photovoltaic cells have an efficiency of 1 %, whereas silicon devices have an efficiency of 15%.

Chapter 13 Formulas

(13-1) $\lambda = 300{,}000{,}000 \,/\, f$ Wavelength in meters.
(13-2) $e = hf$ Energy.
(13-3) $V_{RS} = V_{CC} - V_F$ Voltage value of the current-limiting resistor
(13-4) $R_S = V_{RS} \,/\, I_F$ Resistance value of a current-limiting resistor.

Chapter 13 Answers
Examples

13.1. 3 m
13.2. 4000 Å
13.3. 560 Ω
13.4. 95% approx.
13.5. 4.75 µA approx.
13.6. When the LASCR is reverse biased triggering it by a light signal applied to the gate will not cause it to go into conduction.

Self-Examination

13.1

1. light
2. Radiation
3. electromagnetic spectrum
4. transverse
5. 186,000, 300,000,000
6. wavelength
7. shorter
8. 4000,7700
9. red, violet
10. red, violet
11. photons
12. radiometric
13. photometric
14. flux
15. intensity

16. incidence
17. exitance
18. intensity
19. candela, *or* cd
20. lumen
21. lux

13.2

22. radiant
23. panchromatic
24. incandescent
25. forward
26. wavelengths
27. infrared
28. LED
29. 50, 100
30. heat
31. current-limiting
32. seven-segment
33. dot-matrix
34. source

13.3

35. detector
36. photoemissive, photoconductive, photovoltaic (any order)
37. cathode
38. photoconductive
39. Conductivity
40. photoconductive
41. low
42. high
43. photoconductive
44. photoconductive
45. reverse
46. base
47. photovoltaic
48. higher
49. series
50. parallel

Chapter 13 Glossary

electromagnetic spectrum
A graph that describes radiant energy by showing the location of different frequencies.

wavelength
The distance that an electromagnetic or light wave travels in one cycle.

transverse wave
A wave that causes the particles of a medium to vibrate at right angles to the direction in which the wave is moving.

angstrom
A radiometric measuring unit used to describe the wavelength of an electromagnetic wave. An angstrom is 10^{-10} meters.

quantum theory
A theory based on the absorption and emission of discrete amounts of energy.

photometric system
A system of describing or quantifying light associated with the visible part of the frequency spectrum.

radiometric system
A method of describing or quantifying electromagnetic energy that includes both visible and invisible radiation.

radiant flux
The flow rate of radiant energy per unit of time. It is represented by the Greek letter phi (Φ) and is measured in joules per second, or watts.

radiant intensity
The measure of the radiant power per solid angle unit, or watts per steradian.

radiant incidence
A measure of the radiant energy that strikes the surface of a specific area. It is measured in watts per unit area.

steradian (sr)
The unit of measure of a solid angle originating at the center of a sphere with a radius of 1 m subtended by $1\ M^2$ on the surface of the sphere.

radiant exitance
A measure of the radiant power emitted or released by a specific surface. It is measured in watts per square meter or square centimeter.

radiance
A measure of the radiant intensity that is leaving, passing through, or arriving at a specific surface area. It is measured in watts per steradian meter squared and is determined by dividing the radiant energy intensity from a source by the projected area.

intensity
A measure of the amount of energy contained in an electromagnetic wave.

luminous intensity
The amount of light produced by a photometric source.

candela (cd)
The unit of luminous intensity equal to one lumen per steradian.

luminous flux
The rate at which light energy flows per unit time. The unit of luminous flux is the lumen (lm) and is denoted by the Greek letter phi (Φ).

lumen
The unit used to express the amount of light falling on a surface. One lumen falling on a surface area of one square meter producing an illuminance of one lux.

lux
A unit of illuminance equal to one lumen per meter squared.

illuminance
The density of luminous power that it takes to illuminate a surface. The unit of illuminance is the lux.

luminous exitance
A measure of the amount of luminous flux given off or reflected by a surface. It is measured in lumens per square meter and is used to denote the reflecting capability of a specific surface.

panchromatic source
A radiation source whose radiation extends over a very large portion of the optical spectrum, and thus produces a wide range of visible colors.

heterochromatic source
A radiation source that produces a very limited number of different colors.

monochromatic source
A radiation source that radiates energy of only one specific wavelength or in a very narrow part of the spectrum.

light emitting diode (LED)
A semiconductor optoelectronic device that contains a P-N junction that emits light when forward biased.

infrared emitting diode (IRED)
A type of LED that uses GaAs in its construction and emits infrared energy, which is not visible to the human eye.

photoemissive device
A device that emits electrons in the presence of light.

photoelectric emission
The phenomenon of electrons on the surface of the cathode gaining enough energy to leave the cathode due to the cathode absorbing photons of light.

dynode
The positive plate of a phototube.

photoconductive devices
A device designed so that its resistance decreases when light becomes more intense and increases when light intensity decreases.

Light-activated Silicon Controlled Rectifiers (LASCRs)
A light sensitive SCR which is triggered into conduction when light energy is applied to the gate. It must also have proper voltage values and polarity is applied to the anode and cathode terminals.

photoresistive cell
A device that changes its resistance level and thus its conductivity based on its response to light energy.

dark current
Any conduction current that flows in a photodiode when no light is applied.

photovoltaic devices
A device that converts light energy into electrical energy. When light is applied to a photovoltaic device, the device creates an electrical potential.

photovoltaic cell
convert light energy into electrical energy.

Chapter 14: Integrated Circuits – Chapter Outline

Introduction
Objectives
Key Terms
14.1 Integrated Circuit (IC) Construction
    General Construction
    Monolithic ICs
    MOS ICs
    Film ICs
    IC Packages
    Self-Examination
14.2 Linear ICs
    Operational Amplifiers (Op-Amps)
    Voltage Regulators
    555 Precision Timers
    Self-Examination
14.3 Advantages and Disadvantages of ICs
    Integrated Circuit Limitations
    Self-Examination
Summary
Review Questions
Terms

Chapter 14 Objectives

After studying this chapter, you will be able to:
14.1 Explain the construction differences between various families of integrated circuits.
14.2 Describe the characteristics of linear ICs.
14.3 Based on the advantages and disadvantages of ICs, select to use discrete components or an IC for a given application.
14.4 Follow basic procedures to troubleshoot an IC.

Chapter 14 Key Terms

bipolar IC
bonding pads
diffusion

evaporation process
film IC
metallization
monolithic
MOS IC
thick-film IC
thin-film IC
wafer
yield

Chapter 14 Summary

- Integrated circuits are micro-miniature circuits built on a substrate.
- ICs are made of active and passive components.
- The two general classifications of ICs according to their functions are linear and digital.
- Linear ICs develop an output that is proportional in some way to the input signal; these ICs operate in the straight-line portion of the operating range.
- Monolithic ICs have all components built on a single crystal structure; the components of this IC are all formed on a common substrate at the same time.
- Monolithic ICs that use the same technology as the bipolar transistor in its construction are called bipolar ICs.
- Monolithic ICs that use MOSFETs in their construction are called MOS ICs.
- Film ICs are an assembly of components formed on an insulating material substrate and are classified as hybrid devices.
- Conductors and resistors of a thin-film IC are formed by the evaporation process and are located in a specific area by a screen-printed deposition process; transistors, diodes, capacitors, and MOSFETs are added to the circuit in discrete form.
- Thick-film ICs are similar in many respects to the thin-film ICs; the primary difference is the thickness of the film and the method of material deposition on the film.
- ICs are housed in a number of unique packages; where small size is important for an IC, the flat pack is used.
- A number of packages have been developed to accommodate the reduced size of ICs and to increase the number of pin connections.
- Linear ICs are primarily used to achieve amplification.

- An op amp is a linear IC that has high gain capabilities.
- A positive sign (+) on the op amp denotes the noninverting input; and a negative sign (-) denotes the inverting input.
- A three-terminal voltage regulator is a linear IC.
- Operation of a voltage regulator IC is based on the applied input voltage and the developed output.
- The output voltage of a voltage regulator IC is compared with a reference voltage by a high gain error amplifier; the output of the error amplifier then controls conduction of the series-pass transistor.
- This series-pass transistor of a voltage regulator IC responds as a variable resistor; it changes inversely with the load current to maintain the output voltage at a constant value.
- A 555 timer contains a three-resistor voltage divider, two comparators, a flip flop, an output, a reset transistor, and a discharge transistor.
- Operation of the 555 timer is based on voltage values applied to the trigger and threshold inputs.
- A 555 has two general modes of operation: astable and monostable.
- Astable operation is used in the generation of square or rectangular waveforms.
- Monostable operation is used to produce a wave that can be used to achieve delay or interval timing operations.
- The advantages of ICs are small size, cost, reliability, low power operation, and ease of maintenance.
- The disadvantages of ICs are low power, low voltage, and limited component selection.

Chapter 14 Answers

Self-Examination

14.1

1. substrate
2. micro-miniature
3. wafer
4. chip
5. yield
6. monolithic
7. evaporation
8. Diffusion

9. metallization
10. bipolar
11. MOS
12. insulating
13. dual in-line package, *or* DIP
14. Hybrid
15. MOS
16. resistor

14.2

17. Linear
18. Op-amps
19. high
20. inverting, noninverting (any order)
21. overload
22. linear
23. internal
24. astable, monostable (any order)

14.3

25. advantages
26. reliability
27. discarded
28. power
29. voltage
30. information, data, *or* signals
31. 5, 30 diodes, transistors (any order)

Chapter 14 Glossary

wafer
A piece of circular silicon that serves as the substrate of an IC.

yield
A percentage ratio of the number of good devices produced to the maximum number of possible devices that can be produced in a production run.

monolithic
An IC construction procedure in which all circuit elements are formed and interconnected on or within a single piece of silicon.

evaporation process
A fabrication process in which materials are vaporized and deposited on the surface of another material. Epitaxial growth is an evaporation process.

diffusion
A process by which the atoms of one material are absorbed or moved into another material when subjected to a controlled atmosphere.

bipolar IC
An IC that uses the same technology that is used in the construction of a bipolar transistor.

metallization
The process of attaching metal contacts to the P-type material without permitting the metal to diffuse into the N-type or P-type material.

bonding pads
The points at which leads are attached to the IC.

MOS IC
An IC made of MOSFET technology. MOS ICs use three basic components in their construction: MOSFETs, MOS capacitors, and MOS resistors. They do not require isolation between the source-drain and other P-N junctions formed on the substrate.

film IC
An IC with an assembly of components formed on the surface of an insulating substrate. Conductors and resistors are formed by evaporation and by screen-printed deposition areas. Transistors, diodes, and MOS capacitors are added to the circuit in discrete form and are attached to the circuit with special cement.

thin-film IC
A hybrid assembly in which passive circuit elements and interconnections are formed by evaporation onto a substrate. Active devices are added as discrete components.

thick-film IC
A hybrid IC in which the passive circuit parts and interconnections are formed on an insulating material substrate by print screening. The active components are added in chip form.

Chapter 15: Operational Amplifiers – Chapter Outline

Introduction
Objectives
Key Terms
15.1 Introduction to the Op-Amp
    Op-amp Schematic Symbol
    Basic Op-amp Operation
    Self-Examination
15.2 Inside the Op-Amp
    Internal Circuitry
    Differential Amplifier Stage
    Intermediate Amplifier Stage
    Output Stage
    Self-Examination
15.3 Op-Amp Characteristics
    Open Loop Gain
    Closed Loop Gain
    Common-Mode Rejection Ratio
    Input Offset Voltage
    Input Bias Current
    Input Impedance
    Input Current
    Slew Rate
    Frequency Response
    Bandwidth
    Output Short Circuit Current
    Output Impedance
    Self-Examination
15.4 Analysis and Troubleshooting—Operational Amplifiers
Summary
Formulas
Review Questions
Terms

Chapter 15 Objectives

After studying this chapter, you will be able to:
15.1 Describe the basic operation of an op-amp.

15.2 Describe the function of each op-amp stage.
15.3 Evaluate the performance of an op-amp.
15.4 Analyze and troubleshoot op-amps.

Chapter 15 Key Terms

closed-loop gain
common mode
common-mode rejection ratio (CMRR)
differential amplifier
differential mode
floating state
input bias current
input offset voltage
intermediate amplifier stage
inverting input
noninverting input
open-loop gain
output stage
single-ended mode
slew rate

Chapter 15 – Operational Amplifiers (Op-Amps) – Figure List

Figure 15-1. Op-amp schematic symbol.
Figure 15-2. Op-amp packages.
Figure 15-3. Simplified differential amplifier and equivalent op-amp symbol.
Figure 15-4. Common-mode operation.
Figure 15-5. Differential-mode operation. NEW
Figure 15-6. Single-ended mode operation, noninverting. NEW
Figure 15-7. Single-ended mode operation, inverting. NEW
Figure 15-8. Op-amp diagram depicting the stages of amplification. 3-1, UCOA
Figure 15-9. Internal circuitry of a general-purpose op-amp. 3-6, UCOA
Figure 15-10. Simplified differential amplifier. 3-2, UCOA
Figure 15-11. Ac differential amplifier. 3-3, UCOA
Figure 15-12. Differential amplifier with constant-current source in the emitter. 3-4, UCOA

Chapter 15 Summary

- The minimum number of terminals an op-amp has is five; these consist of two for the power supply voltage, two for the differential input, and one for the output.
- The input terminal labeled – is the inverting input, and the input terminal labeled + is the noninverting input.
- A signal applied to the inverting input is inverted 180° at the output.
- Op-amps are typically supplied voltage by a split power supply.
- The fundamental operation of an op-amp is based on the differential amplifier.
- When two signals of identical values are applied simultaneously to the op-amp inputs, no output signal is produced; this is referred to as the common mode of operation.
- When different signals are applied simultaneously to the op-amp inputs, the output is the conduction difference in Q_1 and Q_2; this is referred to as the differential mode of operation.
- In the single-ended mode of operation, a signal is applied only to one of the inputs while the other input is grounded.
- A signal applied only to the inverting input of a differential amplifier will be amplified and inverted.
- A signal applied only to the noninverting input of a differential amplifier will be amplified without inversion.
- The internal circuitry of an op-amp consists of three functional units or stages of amplification: differential amplifier, intermediate amplifier, and output.

- The differential amplifier stage provides high gain of the signal difference supplied to the inputs and low gain of common signals supplied to the inputs.
- The constant-current source of a differential amplifier provides for low common-mode gain and good signal coupling.
- A differential amplifier can be made with other transistor configurations or devices, such as with a placeDarlington pair or JFETs.
- The intermediate amplifier stage provides for additional gain and dc voltage stabilization.
- The low-impedance output stage develops the power needed to drive an external load.
- The output stage of an op-amp can be single-ended or complementary-symmetry.
- Since a single-ended configuration consumes a great deal of power at high current levels, a complementary-symmetry configuration is typically used.
- Open-loop gain refers to the gain of an amplifier connected to a load that does not have a feedback path between the output and input.
- Closed-loop gain refers to the gain of an amplifier that has a feedback path between the output and input.
- When the output voltage of an open loop op-amp reaches 90% of the source voltage, the op-amp is saturated.
- An input offset voltage is used to overcome the unwanted output of an op-amp.
- Input bias current is the average current flowing into or out of the two op-amp inputs.
- Input impedance is the equivalent resistance that an input source sees when connected to the inputs of an op-amp.
- Slew rate is the rate at which the output of an op-amp changes from one voltage to another in a given time and is an indication of how an op-amp responds to different frequencies.

Chapter 15 Formulas

(15-1) $A_{Vol} = V_{out} / V_{in}$ Open-loop voltage gain.

(15-2) $V_{sat} = 0.90 \times V_{CC}$ Op-amp saturation voltage.

(15-3) $V_{in} = V_{sat} / A_{Vol}$ Input voltage to cause op-amp saturation.

(15-4) $R_2 = (R_1 \times R_F) / (R_1 + R_F)$ Value of R_2 in an op-amp feedback network.

(15-5) $A_V = (R_1 + R_F) / R_1$ Gain in a noninverting op-amp feedback network.

(15-6) $A_V = R_F / R_1$ Gain in an inverting op-amp feedback network.

(15-7) $SR = \Delta V_{out} / \Delta_t$ Slew rate.

(15-8) $f_{max} = SR / (6.28 \times V_{out})$ Maximum frequency before output is distorted.

Chapter 15 Answers

Examples

15-1. 16,667

15-2. 30

15-3. 400,000

15-4. 26.87 V/µS

Self-Examination

15.1

1. amplification
2. +V, −V, inverting input, noninverting input, output (any order)
3. common
4. differential
5. single-ended

15.2

6. differential amplifier, intermediate amplifier, output (any order)
7. differential amplifier
8. common-mode rejection ratio
9. intermediate amplifier
10. single-ended, complementary-symmetry (any order)

15.3

11. open-loop
12. open-loop voltage gain
13. 54 µA
14. negative
15. 40
16. −5
17. input offset voltage
18. slew rate

Chapter 15 Glossary

inverting input
One of two inputs of an operational amplifier or voltage comparator. The inverting input changes the signal 180° in the output and is identified by a – sign on an amplifier symbol.

noninverting input
One of two inputs of an operational amplifier or voltage comparator. This input accepts a signal and causes it to appear in the output without a change in phase. It is identified by a + sign on an amplifier symbol.

differential amplifier
An amplifier having a high common-mode rejection capability and output that is proportional to the difference of its two input signals. This type of amplifier is also called a *difference amplifier*.

common mode
When signals identical in phase and amplitude appear at the inputs of an op-amp simultaneously.

differential mode
When different signals are applied to the inputs of an op-amp simultaneously.

single-ended mode
When a signal is applied to one input of an op-amp and the other end is grounded.

intermediate amplifier stage
The stage of an op-amp that is primarily responsible for additional gain and dc voltage stabilization.

output stage
The stage of an op-amp that has rather low output impedance and is responsible for developing the current needed to drive an external load.

common-mode rejection ratio (CMRR)
The ability of a differential amplifier to cancel a common mode signal.

open-loop gain
The gain of an amplifier connected to a load that does not have a feedback path between the output and input.

closed-loop gain
Gain or amplification achieved by an amplifier that has negative feedback between the input and output.

input offset voltage
Op-amp output voltage that occurs when no differential input voltage is applied.

input bias current
The average current flowing into or out of the two inputs.

slew rate
The rate at which an op-amp's output changes from one voltage to another in a given time. It is expressed in volts per microsecond.

Chapter 16: Linear Op-Amp Circuits – Chapter Outline

Introduction
Objectives
Key Terms
16.1 Inverting Amplifiers
    Self-Examination
16.2 Non-Inverting Amplifiers
    Self-Examination
16.3 Summing Amplifiers
    Self-Examination
16.4 Analysis and Troubleshooting – Linear OP-Amps
Summary
Formulas
Review Questions
Terms

Chapter 16 Objectives

After studying this chapter, you will be able to:
16.1 Describe the operation of an inverting op-amp circuit.
16.2 Describe the operation of a noninverting op-amp circuit.
16.3 Evaluate the operation of a summing op-amp circuit.
16.4 Analyze and troubleshoot linear op-amps circuits

Chapter 16 Key Terms

averaging amplifier
feedback resistor

input resistor
inverting amplifier
noninverting amplifier
scaling adder
subtractor
summing amplifier
virtual ground
voltage gain

Chapter 16 – Linear Op-Amp Circuits – Figure List

Figure 16-1. Inverting op amp.
Figure 16-2. Virtual ground point of an inverting op amp. A−Inverting op-amp circuit. B−Equivalent circuit.
Figure 16-3. Noninverting op amp circuit.
Figure 16-4. Op-amp summing operation.
Figure 16-5. Summing amplifiers. A−Standard adder. B−Adder with gain. C−Scaling adder.

Chapter 16 Summary

- The inverting input of an op-amp produces an output voltage with a $180°$ phase difference from input to output.
- The noninverting input of an op-amp produces an output voltage with a no phase difference between input and output voltages.
- Inverting amplifier voltage gain $(A_v) = R_f/R_{in}$.
- Noninverting amplifier gain $(A_v) = (R_f/R_{in}) + 1$.
- A unity follower is a modified noninverting amplifier with a gain of 1.
- A summing op-amp circuit can add input signal voltages algebraically.
- A subtractor op-amp circuit can be used to determine the algebraic difference between two input signals.

Chapter 16 Formulas

Inverting Op-Amp Circuits

(16-1) $I_{in} = V_{in} / R_{in}$ Inverting input current.
(16-2) $V_{RF} = I_{RF} \times R_F$ Feedback resistor voltage.

Noninverting Op-Amp Circuits

(16-3) $I_{R1} = V_{in} / R_1$ Current through resistor 1.
(16-4) $I_{R1} = V_1 / R_1$ Current through resistor 1.
(16-5) $V_{out} = I_{R1} (R_1 + R_F)$ Output voltage.
(16-6) $A_V = (R_1 + R_F) / R_1$ Gain.
(16-7) $A_V = (R_F / R_1) + 1$ Gain.

Summing Op-Amp Circuits

(16-8) $V_{out} = -(V_1 + V_2)$ Output voltage of a standard adder.
(16-9) $V_{out} = -10(V_1 + V_2 + V_3)$ Output voltage of an adder with gain.
(16-10) $V_{out} = -(V_1 \times (R_F / R_1) + V_2 \times (R_F / R_1) + V_3 \times (R_F / R_3) + V_4 \times (R_F / R_4))$ Output voltage of a scaling adder.

Chapter 16 Answers

1. Two or more
2. 0V
3. 180°
4. 0°
5. The voltage difference between input

Examples

16-1. 2.5V
16-2. A_v = -20, Z_{in} = 10 KΩ
16-3. 2000 kΩ or 2 M Ω
16-4. 134.33
16-5. +4 V
16-6. −10.75

Self-Examination

16.1

1. inverting
2. virtual ground
3. −5

4. $-1.55\ V_{pp}$
5. R_{in}

16.2

6. noninverting amplifier
7. inverting
8. 40
9. 7.8 V
10. high

16.3

11. addition
12. -4
13. voltage gain
14. unity
15. difference

Chapter 16 Glossary

inverting amplifier
An op-amp circuit that receives a signal voltage at its input and delivers a large undistorted version of the signal at its output.

virtual ground
The point of a circuit that is at zero potential (0 V) but is not actually connected to ground.

noninverting amplifier
An op-amp circuit that provides controlled voltage gain with high input impedance and no inversion of the input-output signals.

Unity follower
A modified noninverting amplifier circuit which has a gain of 1, by shorting out the feedback resistor and removing the input resistor. In this circuit so that the output is identical to the input.

summing point
A common point that isolates input voltage values applied to the inverting input.

scaling adder
A summing amplifier that has different weighted values applied to its input resistors.

Chapter 17: Specialized Op-Amp Circuits – Chapter Outline

Introduction
Objectives
Key Terms
17.1 Controlled Voltage and Current Sources
    Self-Examination
17.2 Converters
    Self-Examination
17.3 Comparators
    Self-Examination
17.4 Integrators
    Self-Examination
17.5 Differentiators
    Self-Examination
17.6 Precision Rectifiers
    Self-Examination
17.7 Instrumentation Amplifiers
    Self-Examination
17.8 Analysis and Troubleshooting – Specialized OP-Amp Circuits
Summary
Formulas
Review Questions
Terms

Chapter 17 Objectives

After studying this chapter, you will be able to:
17.1 Explain the operation of a controlled voltage and current source.
17.2 Analyze the operation of a converter op-amp.
17.3. Explain the operation of a comparator.
17.4 Describe the operation of an integrator.
17.5 Explain how a differentiator works.
17.6 Analyze the operation of a precision rectifier
17.7 Analyze the operation of an instrumentation op-amp

Chapter 17 Key Terms

constant-current sources
constant-voltage source

voltage-controlled voltage source
voltage-controlled constant current source
converters
voltage-to-current converters
current-to-voltage converters
comparators
adjustable-referenced comparator
active circuit
waveshaper
integrators
differentiator
precision rectifiers
external gain resistor
instrumentation amplifier

Chapter 17 – Specialized Op-Amp Circuits – Figure List

Figure 17-1. Constant voltage source.
Figure 17-2. Inverting constant current source.
Figure 17-3. Current-to-voltage converter.
Figure 17-4. Voltage-to-current converter.
Figure 17-5. Zero-referenced inverting comparator.
Figure 17-6. Zero-referenced non-inverting comparator.
Figure 17-7. Adjustable-referenced comparator.
Figure 17-8. A basic integrator circuit.
Figure 17-9. An op-amp integrator.
Figure 17-10. An op-amp differentiator.
Figure 17-11. Silicon diode rectifier.
Figure 17-12. Precision rectifier.
Figure 17-13. Internal construction of an instrumentation amplifier.
Figure 17-14. Low cost, low power instrumentation amplifier, AD620 (Courtesy Analog Devices).

Chapter 17 Summary

- A voltage controlled voltage source generates a constant voltage regardless of the electrical load. The voltage value depends on a controlling voltage source circuit.

- A voltage controlled constant source generates a constant current regardless of the electrical load. The current value depends on a controlling voltage source circuit.
- A voltage to current converter transforms the electrical behavior of the system. The input voltage controls the output voltage in such a way that the output current remains constant.
- A current to voltage converter transforms the electrical behavior of the system. The input voltage controls the output current in such a way that the output voltage remains constant.
- In an op-amp comparator circuit the values of the voltage inputs applied to its inverting and non-inverting input terminals is compared. It can generate a very high output voltage even when there is a minute voltage difference between the applied inputs.
- An op-amp integrator is a waveshaping circuit which generates an output voltage corresponding to the addition of specific quantities of the input signal over a period of time.
- A differentiator is a waveshaping circuit which develops an output voltage corresponding to the rate of change of the applied input. This operation is the inverse of an integrator.
- A precision op-amp/diode rectifier circuit is used for measuring and detecting minute ac input voltages.
- An instrumentation amplifier generally consists of at least three op-amps that are connected in such as way as to minimize noise and have high gain capabilities.

Chapter 17 Formulas

(17-1) $A_{cl} = R_f/R_{in}$
(17-2) $V_{out} = A_{cl} \times V_{in}$
(17-3) $I_{in} = V_{in}/R_{in}$
(17-4) $V_{out} = I_{in} \times R_f$
(17-5) $V_{diff} = V_{in} - V_f$
(17-6) $V_{RL} = I_{out} \times R_L$
(17-7) $V_{Rf} = I_{out} \times R_f$
(17-8) $V_{out} = V_{in} \times A_{ol}$
(17-9) $t = 5\,RC$
(17-10) $A_{CL} = 1 + \frac{2R}{R_G}$
(17-11) $R_G = \frac{2R}{A_{CL} - 1}$

Chapter 17 Answers
Examples

17-1. $A_{cl}=5$, $I_{in}=0.8$ mA , $I_f= 0.8$ mA, $V_{out}=4$ V

17-2. $I_L=2$ mA; I_L remains unchanged

17-3. $V_{out} = 4$ V

17-4. $I_{out} = 10$ mA, $V_{sat}= 10$ V, $V_{Rf} = 0.5$ V, $V_{diff}=0.45$ V

17-5. On the application of an ac sine wave input, with $V_{p-p}=0.15$mV, the output will be a square waveform, with $V_{p-p}=30$ V. The output waveform is 180° out of phase with the applied ac input signal.

17-6. On the application of an ac sine wave input, with $V_{p-p}=0.15$mV, the output will be a square waveform, with $V_{p-p}=30$ V. The output waveform is in phase with the applied ac input signal.

17-7. $V_{out} = 5$ V, $V_{diff}= 0.05$mV

17-8. 5 ms

17-9. $A_{CL}=201$

17-10. $R_G≈134$ Ω

Self-Examination

17.1

1. (b) constant-value of current
2. (b) depends on the value of the controlling voltage
3. (a) V_{out}
4. (a) the load resistor
5. (c) remain unchanged

17.2

6. (a) changes one form of energy into another form of energy
7. (a) the load and feedback resistor in series
8. (c) a feedback resistor and no input resistor.
9. voltage, current
10. current, voltage

17.3

11. (d) open circuit
12. (c) making voltage comparisons
13. (b) negative supply voltage
14. (a) open loop configuration
15. (d) extremely high

17.4

16. (c) capacitor
17. (a) waveshaping
18. (d) triangular wave
19. (a) constant
20. (a) constant

17.5

21. (a) resistor
22. (a) waveshaping
23. (d) square wave
24. proportional to the rate of change of the input signal
25. (a) zero

17.6

26. (b) very low diode forward bias voltages
27. (b) a diode connected directly to the output of the op-amp
28. (b) 0 to 0.7 V
29. (b) half-wave rectified signal
30. (b) precision rectifier circuits containing op-amps and diodes

17.7

31. (a) 0 V
32. (c) differential amplifier
33. (d) three op-amps and seven resistors
34. (b) setting the voltage gain
35. (a) high noise environment

Chapter 17 Glossary

voltage controlled voltage source
An electronic circuit which generates a constant voltage regardless of the electrical load. The voltage value depends on a controlling voltage source circuit.

voltage controlled current source
An electronic circuit which generates a constant current regardless of the electrical load. The current value depends on a controlling voltage source circuit.

converters
An electronic circuit which transforms the electrical behavior of the system

voltage to current converter
An electronic circuit in which the input voltage controls the output voltage in such a way that the output current remains constant.

current to voltage converter
An electronic circuit in which the input voltage controls the output current in such a way that the output voltage remains constant.

comparator
An op-amp function that compares the voltage applied to one input, to that applied to another input, that is of a predetermined value or reference.

waveshaper
An electronic circuit which alters the shape of the waveform over a given time duration.

active circuit
An electronic circuit which has passive components such as resistors and capacitors, connected to amplifying or control devices such as op-amps and transistors.

integrator
An op-amp circuit used for generating output voltage corresponding to the addition of specific quantities of the input signal over a period of time.

differentiator
An op-amp circuit used for developing an output voltage corresponding to the rate of change of the applied input signal.

precision rectifier
An op-amp/diode circuit that has the ability to conduct at very low diode forward bias voltages.

external gain resistor
The resistor used to set the gain of an instrumentation amplifier.

instrumentation amplifier
An instrumentation amplifier has high gain, high common mode rejection and is used for amplifying small signals in an environment which has electrical noise.

Chapter 18: Voltage Regulator Circuits – Chapter Outline

Introduction
Objectives
Key Terms
18.1 Voltage Regulation
    Line Regulation
    Load Regulation
    Self-Examination
18.2 Linear Voltage Regulators –Series and Shunt
    Series Linear Voltage Regulators with Current Limiting Protection
    Foldback Current Limiting in Series Voltage Regulators
    Shunt Linear Voltage Regulators
    Over-Voltage Protection in Shunt Linear Voltage Regulators
    Self-Examination
18.3 Switching Voltage Regulators—Step-Down and Step-Up
    Step Down Switching Regulators
    Step Up Switching Regulators
    Self-Examination
18.4 IC Voltage Regulators
    Fixed Linear IC Voltage Regulators
    Adjustable Linear IC Voltage Regulators
    Switching IC Voltage Regulators
    Self-Examination
18.5 Analysis and Troubleshooting– Voltage Regulators
Summary
Formulas
Review Questions
Problems
Terms

Chapter 18 Objectives

After studying this chapter, you will be able to:
18.1 Explain the basic concept of voltage regulation
18.2 Describe how a linear voltage regulator responds to input/output changes
18.3 Describe how a switching voltage regulator responds to input/output changes
18.4 Interpret the datasheet of an IC voltage regulator
18.5 Analyze and troubleshoot voltage regulator circuits

Chapter 18 Key Terms

Regulator
Line regulator
Load regulator
Thermal overload
Linear regulator
Switching regulator
Series regulator
Shunt regulator
Step-up regulator
Step-down regulator
Foldback regulation
Pulse width modulation

Chapter 18 – Voltage Regulator Circuits – Figure List

Figure 18-1. Basic power supply.
Figure 18-2. Regulated power supply.
Figure 18-3. Line regulation.
Figure 18-4. Load regulation.
Figure 18-5. Block diagram of a series regulator.
Figure 18-6. Series voltage regulator circuit. NEW
Figure 18-7. Foldback limiting in an overloaded voltage regulator. NEW
Figure 18-8. Block diagram of a shunt regulator. NEW
Figure 18-9. Shunt voltage regulator. NEW
Figure 18-10. Block diagram of a step-down switching regulator. NEW
Figure 18-11. Step-down switching voltage regulator. NEW
Figure 18-12. Switching regulator waveforms. NEW
Figure 18-13. Step-up switching voltage regulator. NEW
Figure 18-14. 7800/7900 voltage regulators. NEW
Figure 18-15. Datasheet for 78XX voltage regulator. NEW
Figure 18-16. Circuit diagrams of adjustable voltage regulators. NEW
Figure 18-17. Datasheet for LM317 positive adjustable voltage regulator. pdf
Figure 18-18. Datasheet for LM337 negative adjustable voltage regulator. pdf
Figure 18-19. Switching IC voltage regulator.
Figure 18-20. Step-down switching regulator.

Chapter 18 Summary

- Voltage regulators operating within specified conditions maintain a constant output voltage regardless of changes in either the input voltage or its load current demand.
- Voltage regulators are generally described as being linear or switching based on whether the current flow to the load device is continuous or interrupted during operation respectively.
- The ability of a regulator to maintain an output voltage at a constant value regardless of variation in the input voltage is called line regulation.
- The ability of a regulator to maintain an output voltage at a constant value regardless of variations in the load current demand is called load regulation.
- Linear regulators are further classified as being either series or a shunt (parallel), based on how certain electronic components are connected to the load device.
- Most series linear voltage regulator circuits employ some form of overload protection to prevent the series control element from being damaged, whereas most shunt linear voltage regulators require protection from input over-voltage conditions.
- Switching regulators are classified as being either step-down or step-up. When the switching electronic component is in series with the load it will reduce or step-down the voltage. It steps-up the voltage when the switching electronic component is in parallel with the load.
- The efficiency of a voltage regulator can be improved is by operating the control element as a switching device by rapidly switching it on and off.
- A pulse width modulator (PWM) circuit is frequently employed in switching the control element of a regulator. It typically generates a pulsating DC waveform in which the width of the pulses can be adjusted.
- ICs voltage regulators have the essential regulator components built on a single chip instead of using discrete components.
- The 7800 series is a representative type of fixed positive voltage regulators, and the 7900 series for fixed negative voltage regulators.
- Adjustable IC voltage regulators are used in applications where a non-standard regulator voltage values are needed.

Chapter 18 Formulas

(18-1) Line regulation=$\frac{\Delta V_{out}}{\Delta V_{in}}$

(18-2) Load regulation=$\frac{\Delta V_{out}}{\Delta I_L} = \frac{V_{No\ Load}-V_{Full\ Load}}{I_{Full\ Load}-I_{No\ Load}}$

(18-3) Zener current: $I_Z = \frac{V_{in} - V_Z}{R_S}$

(18-4) Voltage output of developed by an op-amp when comparing sample voltage using resistor network R_1, R_2 with a Zener voltage: $V_{out} = V_Z \times (1 + \frac{R_1}{R_2})$

(18-5) Maximum load current in series voltage regulator with current limiting: $I_{L(max)} = \frac{0.7}{R_{SC}}$

(18-6) % Duty Cycle=$\frac{t_{on}}{T} \times 100 = \frac{t_{on}}{t_{on} + t_{off}} \times 100$

(18-7) Average value of output voltage for a square wave pulsating DC waveform: $V_{out} = \frac{t_{on}}{T} \times V_{in}$

(18-8) Output voltage developed by the adjustable voltage regulator LM317:

$$V_{out} = V_{REF} \times \left(1 + \frac{R_2}{R_1}\right) + I_{ADJ} \times R_2$$

Chapter 18 Answers

Examples

18-1. 3.63μV/V

18-2. 0.0093333V/mA = 9.33mV/mA

18-3. Regulator I would be better since the output voltage under full load is 9.9985V, whereas that for regulator II is 9.95V.

18-4. R_2=30kΩ, R_S=320Ω

18-5. R_{SC}=5.6Ω

18-6. f=100Hz, % duty cycle=25%, V_{out}=3.5V

18-7. V_{out} range: 1.25V – 20.42V

Self-Examination

18.1

1. remain constant
2. increase
3. remain constant
4. increase
5. Increase voltage, increase current demand

18.2

6. series
7. increases
8. decreases

9. parallel
10. increase

18.3

11. alternating between cutoff and saturation
12. $t_{on} = t_{off}$
13. series
14. shunt
15. increases

18.4

16. 3, positive
17. heat sinks
18. external
19. duty cycle
20. Time period

Chapter 18 Glossary

Line Regulation
The ability of a regulator to maintain an output voltage at a constant value regardless of variations in the input voltage is called line regulation.

Load Regulation
The ability of a regulator to maintain an output voltage at a constant value regardless of variations in the load current demand is called load regulation.

Linear voltage regulator
A voltage regulator in which the control element, usually a transistor, is operated in the linear region, and is never completely switched off and on during its operation.

Switching voltage regulator
A voltage regulator in which the control element, usually a transistor, is completely switched off and on as part of its regulation normal operations.

Series linear voltage regulator
A voltage regulator in which the control element is connected in series with the load device and operates in its' linear region.

Shunt linear voltage regulator
A voltage regulator in which a control element is connected in shunt or parallel with the load device and operates in its' linear region.

Foldback current limiting
A current limiting circuit in a linear voltage regulator which is used for reducing the amount of current flow if the load exceeds the maximum permitted current flow through the load.

Step-up switching voltage regulators
A type of switching regulator in which the value of the output voltage (V_{out}) generated is smaller than or equal to the applied input voltage (V_{in}).

Step-down switching voltage regulators
A type of switching regulator in which the value of the output voltage (V_{out}) can be increased beyond the applied input voltage (V_{in}).

Pulse Width Modulation (PWM)
The process of generates a pulsating DC waveform in which the width of the pulses can be adjusted.

Duty Cycle
The ratio of the on time (t_{on}) to the time period (T) of a pulsating DC waveform usually expressed as a percentage.

Fixed IC voltage regulator
An integrated circuit voltage regulator which generates a fixed positive or negative voltage output when operated within rated specifications.

Adjustable IC voltage regulator
An integrated circuit voltage regulator which generates a variable positive or negative voltage output when operated within rated specifications. The voltage is usually varied by adjusting the values of externally connected components.

Chapter 19: Filter Circuits – Chapter Outline

Chapter 19 Objectives

After studying this chapter, you will be able to:
19.1 Explain the operation of low-pass, high-pass, band-pass and band-stop filter circuits
19.2 Plot frequency response curves for low-pass, high-pass, band-pass filter, and band-stop filter circuits
19.3 Explain the use of decibels to measure voltage and power gain in filter circuits
19.4 Analyze the operation of resonant circuits
19.5 Explain the operation of an active filter circuit
19.6 Analyze and troubleshoot filter circuits

Chapter 19 Key Terms

attenuation
bandwidth
low-pass filter

high-pass filter
band-pass filter
frequency response
quality factor (Q)
power gain/loss
power-loss ratio
voltage-loss ratio
multiple-order filters
center frequency
3 dB-down frequency
half-power frequency
resonant frequency
resonant circuit
active filter
dB/decade
dB/octave
selectivity

Chapter 19 Summary

- A low-pass filter passes low-frequency signals up to a specified cutoff frequency and blocks high-frequency signals beyond the cutoff.
- A high-pass filter passes high-frequency signals beyond a specified cutoff frequency and blocks low-frequency signals below the cutoff.
- A band-pass filter is a combination of high- and low-pass filter sections that will pass a mid-range of frequency signals and blocks other frequencies.
- A band-stop or notch filter blocks a mid-range of frequency signals and passes other frequencies.
- The performance of a filter circuit can be evaluated by using frequency response curves, which plots the power and voltage gain/loss in decibels (dB) over a range of frequencies.
- The 0 dB region of a frequency response curve signifies the portion where the output power is equal to the input power, so the signal passes through without any loss.
- The 3 dB point(s) on a frequency response curve signify the point(s) where the output power has been reduced to half the value of the input power.
- The bandwidth of a filter is the frequency range in which the output power is greater than half the value of the input power.

- The quality factor (Q) of a filter circuit refers to the ratio of the resistance and the reactance. The bandwidth of a filter and its quality factor are inversely related.
- Resonant circuits are designed to pass a certain range of certain frequencies and to block other frequencies. Resonant circuits use resistors, inductors and capacitors in their construction, which may be connected in series or parallel.
- Active filters use amplifying components such as op-amps, in addition to passive components such as resistors, inductors and capacitors. They are designed to pass a specific range of frequencies while blocking others.
- The order of a filter is based on the number of cascaded filter sections.
- As the order number of a filter increases the loss in dB per decade beyond the 3 dB point increases linearly by multiples of 20 dB per decade.

Chapter 19 Formulas

(19-1) $X_C = \frac{1}{2\pi f_c C} = R$

(19-2) $f_c = \frac{1}{2\pi RC}$

(19-3) $X_L = 2\pi f_c L = R$

(19-4) $f_c = \frac{R}{2\pi L}$

(19-5) $X_C = \frac{1}{2\pi f_c C} = R$

(19-6) $f_c = \frac{1}{2\pi RC}$

(19-7) $X_L = 2\pi f_c L = R$

(19-8) $f_c = \frac{R}{2\pi L}$

(19-9) $\text{BW} = f_{cu} - f_{cl}$

(19-10) $f_{cf} = \sqrt{f_{cu} \times f_{cl}}$

(19-11) $Q = \frac{f_{cf}}{f_{cu} - f_{cl}}$

(19-12) $\text{BW} = f_{cu} - f_{cl}$

(19-13) $f_{cf} = \sqrt{f_{cu} \times f_{cl}}$

(19-14) Power gain (dB) $= 10 \log_{10} \frac{P_{out}}{P_{in}}$; when $P_{out} > P_{in}$

(19-15) Power loss (dB) $= -10 \log_{10} \frac{P_{out}}{P_{in}}$; when $P_{out} < P_{in}$

(19-16) Voltage gain (dB) $= 20 \log_{10} \frac{V_{out}}{V_{in}}$

(19-17) $\frac{P_{out}}{P_{in}} = 10^{\frac{dB}{10}}$

(19-18) $\frac{V_{out}}{V_{in}} = 10^{\frac{dB}{20}}$

(19-19) $f_r = \frac{1}{2\pi\sqrt{LC}}$

(19-20) $Q = X_L/R$

(19-21) $BW = f_r/Q$

(19-22) $f_{cf} = \frac{1}{2\pi C} \sqrt{\frac{R_1+R_2}{R_1 R_2 R_3}}$

(19-23) $Q = \pi f_{cf} C R_3$

(19-24) $BW = f_{cf}/Q$

Chapter 19 Answers

Examples

19-1. $X_C = 1\ k\Omega$, $C = 1.59\ \mu F$

19-2. $f_c = 63.69$ Hz

19-3. $f_c = 3.185$ kHz

19-4. $f_{cf} = 2.236$ kHz

19-5. $BW = 100,000$

19-6. $V_{out}/V_{in} = 0.0224$

19-7. $f_r = 45.039$ kHz

19-8. $f_r = 45.039$ kHz; $Q=7.07$; $BW=6.370$ kHz

19-9. $f_c = 63.69$ Hz

Self-Examination

19.1

1. (a) low-pass
2. (b) high-pass
3. (c) band-pass
4. (d) band-stop
5. (b) capacitive, inductive

19.2

6. (b) half
7. (b) half
8. (b) less than 1
9. (c) equal to the input
10. (c) band-pass

19.3

11. (c) $X_L = X_C$
12. (b) minimum
13. (b) small

14. (a) maximum
15. (a) decreases

19.4

16. (d) op-amp
17. (c) cascaded filters
18. (d) low
19. (c) cutoff frequency
20. (a) increase

Chapter 19 Glossary

Amplification
An increase in value

Attenuation
A reduction in value

Band-pass filter
A frequency-sensitive ac circuit that allows incoming frequencies within a certain band to pass through but attenuates frequencies below or above this band.

Decibel (dB)
A unit used to express an increase or decrease in power, voltage, or current in a circuit.

Filter
A circuit used to pass certain frequencies and attenuate all other frequencies.

Frequency
The number of ac cycles per second, measured in hertz (Hz)

Frequency response
A circuit's ability to operate over a range of frequencies.

High-pass filter
A frequency-sensitive ac circuit that passes high frequency input signals to its output and attenuates low-frequency signals.

Low-pass filter
A frequency-sensitive ac circuit that passes low frequency input signals to its output and attenuates high-frequency signals.

Parallel resonant circuit
A circuit that has an inductor and a capacitor connected in parallel that causes it to respond to frequencies applied to the circuit.

Quality factor (Q)
The 'figure of merit' or Q of frequency-sensitive circuit is the ratio of the resistance to the reactance.

Resonant circuit
A frequency-sensitive circuit in which the inductive and capacitive reactance values are used to determine the resonant frequency.

Resonant frequency
The frequency that passes through or is blocked by a resonant circuit depends on the inductive and capacitive reactance.

Selectivity
The ability a resonant or filter circuit to select a specific frequency and reject all other frequencies.

Series resonant circuit
A circuit that has an inductor and a capacitor connected in series that causes it to respond to frequencies applied to the circuit.

Cutoff frequency
The frequency at the -3 dB point, where the pass-band starts to roll off

Order
A term that indicates the number of filter networks that are cascaded. As the order number increases the amount of filtering increases linearly by 20 dB/decade beyond the 3 dB points.

Active filters
Filter circuits which use amplifying components such as op-amps, in addition to passive components such as resistors, inductors and capacitors. They are designed to pass a specific range of frequencies while blocking others.

Decade
A frequency change by a factor of ten.

Octave
A frequency change by a factor of two.

Chapter 20: Oscillator Circuits – Chapter Outline

Introduction
Objectives
Key Terms
20.1 Oscillator Fundamentals
    Oscillator Types
    Feedback Oscillator Operation
    Relaxation Oscillator Operation
    Self-Examination
20.2 Feedback Oscillators
    Armstrong Oscillator
    Hartley Oscillator
    Colpitts Oscillator
    Clapp Oscillator
    Crystal Oscillator
    Pierce Oscillator
    Wien-Bridge Oscillator
    Self-Examination
20.3 Relaxation Oscillators
    UJT Oscillators
    Multivibrators
    Astable Multivibrators
    Monostable Multivibrators
    Bistable Multivibrators
    IC Waveform Generators
    555 Flip-Flop
    555 Astable Multivibrators
    Self-Examination
    Blocking Oscillator
20.4 Analysis and Troubleshooting– Oscillator Circuits
Summary
Formulas
Review Questions
Problems
Terms

Chapter 20 Objectives

After studying this chapter, you will be able to:
20.1 Analyze the operation of LC, RC, and RL circuits.
20.2 Analyze common feedback oscillators.
20.3 Analyze common relaxation oscillators.
20.4 Analyze and troubleshoot oscillator circuits.

Chapter 20 Key Terms

astable multivibrator
capacitance ratio
continuous wave (CW)
damped sine wave
feedback oscillator
flip-flop
free-running multivibrator
monostable multivibrator
multivibrator
nonsymmetrical multivibrator
piezoelectric effect
pulse repetition rate (PRR)
regenerative feedback
relaxation oscillator
symmetrical multivibrator
capacitance ratio
one-shot multivibrator
tank circuit
sharp filter
threshold terminal
triggered multivibrator
upper trip point (UTP)
lower trip point (LTP)

Chapter 20 – Oscillator Circuits – Figure List

Figure 20-1. Fundamental oscillator parts.
Figure 20-2. LC tank circuit being charged. A–Basic circuit. B–Charging action. C–Charged capacitor.

Figure 20-3. Charge-discharge action of an LC circuit.

Figure 20-4. Wave types. A—Damped oscillatory wave. B—Continuous wave.

Figure 20-5. RF oscillator coils. 5-6, UCOA

Figure 20-6. *RC* charging action. A—Circuit. B—Circuit values. 15-15, EE5

Figure 20-7. *RC* discharging action. A—Circuit. B—Circuit values. 15-16, EE5

Figure 20-8. Armstrong oscillator. A—Circuit. B—Characteristic curves. 15-7, EE5

Figure 20-9. Hartley oscillator. 15-8, EE5

Figure 20-10. Colpitts oscillator. 15-9, EE5

Figure 20-11. placeCityCrystal equivalent circuits. A—Series resonant. B—Parallel resonant. 15-10, EE5

Figure 20-12. Crystal-controlled oscillators. A—Hartley. B—Colpitts. 15-11, EE5

Figure 20-13. Pierce oscillator. 15-12, EE5

Figure 20-14. A - Wien-Bridge oscillator. B – Positive frequency response curve. NEW

Figure 20-15. UJT oscillator. 15-17, EE5

Figure 20-16. A - Astable multivibrator circuit. B – Symmetrical waveform. C – Non-symmetrical waveform. 15-18, EE5

Figure 20-17. Monostable multivibrator. 15-19, EE5

Figure 20-18. Monostable multivibrator waveforms. A—Trigger input waveform. B—Differentiator output waveform. C—Monostable multivibrator output waveform. 15-20, EE5

Figure 20-19. A - Bistable multivibrator circuit. B – Input/Output relationships. 15-21, EE5

Figure 20-20. A - Internal block diagram of an LM555 IC. B – Input/Output combinations. 15-22, EE5

Figure 20-21. Astable multivibrator. A – LM555 IC and external components. B – Pin designations. C – Internal block diagram of an astable multivibrator.

Figure 20-22. Astable multivibrator waveforms.

Figure 20-23. NE/SE 555C datasheet (Courtesy Philips Semiconductors).

Chapter 20 Summary

- An oscillator generates a continuously repetitive output signal.
- Two basic types of oscillators based on their method of operation are the feedback oscillator and relaxation oscillator.

- In a feedback oscillator circuit, a portion of the output power is returned to the input circuit.
- A feedback oscillator circuit consists of an amplifier, feedback network, frequency-determining network, and a dc power source.
- A relaxation oscillator responds to an electronic device that goes into conduction for a certain time and then turns off for a period of time.
- Relaxation oscillators usually generate square- or triangular-shaped waves.
- The active device of a relaxation oscillator is triggered into conduction by a change in voltage.
- An inductive-capacitance network usually determines the operating frequency of a feedback oscillator.
- Without regenerative feedback, a tank circuit produces a damped sine wave.
- The operation of a relaxation oscillator depends on the charge and discharge of an RC or RL network.
- Feedback in a feedback oscillator can be accomplished by inductance, capacitance, or resistance coupling.
- The following oscillators are feedback oscillators: Armstrong, Hartley, Colpitts, Clapp, crystal, Pierce, and Wien-bridge.
- An Armstrong oscillator is also referred to as a tickler coil.
- The frequency of an Armstrong oscillator is based on the value of a capacitor and the secondary of a transformer.
- The Hartley oscillator incorporates a tapped coil and is used extensively in AM and FM radio receivers.
- A Colpitts oscillator is similar to the shunt-fed Hartley oscillator except that is uses two capacitor instead of a tapped coil.
- The amount of feedback developed in a Colpitts oscillator is based on the capacitance ratio of the two capacitors in the tank circuit.
- The Clapp oscillator is similar to the Colpitts oscillator and uses an additional capacitor in series with the coil of the tank circuit.
- Crystals oscillators are used when extremely high frequency stability is desired.
- The Hartley and Colpitts oscillators are often modified to accommodate a crystal.
- A crystal has the property to change electrical energy into mechanical energy and mechanical energy into electrical energy; this is known as the piezoelectric effect.
- The Wien-Bridge oscillator uses a RC network for providing positive feedback to an op-amp for controlling the frequency of oscillation, and negative feedback for controlling the gain.

- The Pierce oscillator is a modification of the Colpitts oscillator; it uses a crystal in place of the tank circuit inductor.
- Relaxation oscillators are typically used to generate nonsinusodial waveforms, such as the sawtooth and rectangular waveforms.
- Relaxation oscillators depend on the charge and discharge of a capacitor-resistor network.
- UJT oscillators are used in applications that require a signal with a slow rise time and a rapid fall time.
- A multivibrator is a classification of the relaxation oscillator.
- Two general types of multivibrators are triggered and free-running.
- An astable multivibrator is free-running.
- Monostable and bistable multivibrators are free-running.
- A monostable multivibrator has one stable state of operation; it is often called a one-shot multivibrator.
- A bistable multivibrator has two stable states of operation; it is often called a flip-flop.
- A 555 IC can be configured to respond as an astable multivibrator.
- The internal circuitry of a 555 IC consists of two comparators, a bistable flip-flop, a resistive divider, a discharge transistor, and an output stage.
- The shape of the waveform from an IC astable multivibrator is determined by an external RC network.

Chapter 20 Formulas

(20-1) $A_{cl} = A_v \times B$ Closed loop feedback gain of an oscillator

(20-2) $f_r = \frac{1}{2\pi\sqrt{LC}}$ Resonant frequency of a tank circuit.

(20-3) $f_r = \frac{1}{2\pi\sqrt{L_1 C_1}}$ Frequency of the Hartley oscillator

(20-4) $B = \frac{L_b}{L_a}$ Attenuation factor, B, of the Hartley oscillator

(20-5) $C_T = \frac{C_1 \times C_2}{C_1 + C_2}$ Total capacitance is called C_T of a Colpitts oscillator

(20-6) $f_r = \frac{1}{2\pi\sqrt{L_1 C_T}}$ Frequency of a Colpitts oscillator

(20-7) $B = \frac{C_1}{C_2}$ Attenuation factor B of a Colpitts oscillator

(20-8) $C_T = \frac{1}{\frac{1}{C_1} + \frac{1}{C_2} + \frac{1}{C_3}} = \frac{C_1 C_2 + C_2 C_3 + C_1 C_3}{C_1 C_2 C_3}$ Total capacitance C_T of a Clapp oscillator

(20-9) $f_r = \frac{1}{2\pi\sqrt{L_1 C_3}}$ Frequency of a Clapp oscillator when $C_3 \ll C_1$ or C_2

(20-10) $f_r = \frac{1}{2\pi RC}$ Frequency of a Wien-bridge oscillator, when $R_1 = R_2 = R$ and $C_1 = C_2 = C$

(20-11) $A_{CL} = 1 + \frac{R_4+R_5}{R_3}$ Closed-loop gain of a Wien-bridge oscillator

(20-12) $V_P = \eta V_{BB} + 0.7$ Peak voltage (V_P) of UJT

(20-13) $f = \frac{1}{1.4 \times R \times C}$ Output frequency of a symmetrical multivibrator.

(20-14) $PW = 0.7R_2C_1$ Pulse width of monostable multivibrator

(20-15) UTP $= \frac{2}{3}V_{CC}$ Upper trip point of a 555 IC.

(20-16) LTP $= \frac{1}{3}V_{CC}$ Lower trip point of a 555 IC.

(20-17) $f_{out} = \frac{1}{T}$ Output frequency of an astable multivibrator.

(20-18) $T = 0.693(R_A + 2R_B)C$ Cycle time of an astable multivibrator.

(20-19) $f_{out} = \frac{1}{0.693(R_A+2R_B)C}$ Output frequency of an astable multivibrator.

(20-20) $PW = 0.693(R_A + R_B)C$ Pulse width of an astable multivibrator.

(20-21) % duty cycle $= \frac{On-time}{Total\ cycle\ time} = \frac{R_A+R_B}{R_A+2R_B} \times 100$ Duty cycle of an astable multivibrator.

Chapter 20 Answers

Examples

20.1 5.035 kHz

20.2 t = 0.5 ms, so T = 2.5ms to fully charge the capacitor

20.3 5.035 kHz

20.4 C_T= 0.0578 µF, f_r = 5. 410 kHz

20.5 C_T= 150 pF, f_r = 183.87 kHz

20.6 f_r = 636 Hz

20.7 t_D = 10 µs, T_D = 50 µs (time taken to fully discharge the capacitor)

20.8 t_2-t_1 = 5 ms

20.9 T= 2.079 ms, f_{out}= 481 Hz, PW=1.213 ms, % duty cycle = 58.35%

Self-Examination

20.1

1. oscillator
2. feedback network, amplifier, frequency-determining network, and dc power source (any order)
3. feedback
4. tank
5. resonant
6. damped
7. 4.1 kHz

8. Decreases
9. relaxation
10. feedback

20.2

11. *L, C* (any order)
12. tapped coil
13. capacitors
14. crystal
15. Wien-Bridge

20.3

16. Relaxation
17. relaxation
18. triggered (or monostable or one-shot), free-running (or astable)
19. astable
20. one-shot
21. flip-flop
22. 555 IC
23. 50
24. unsymmetrical
25. cycle time

Chapter 20 Glossary

feedback oscillator
A type of oscillator in which a portion of the output power is returned to the input circuit.

regenerative feedback
Feedback from the output to the input that is in phase so that it is additive.

Armstrong oscillator
An oscillator that uses a transformer as a feedback element to produce a 180° phase shift which is needed for producing oscillation.

relaxation oscillato
A nonsinusodial oscillator that has a resting or nonconductive period during its operation.

tank circuit
A parallel resonant LC circuit.

damped sine wave
A wave in which successive oscillations decrease in amplitude.

continuous wave (CW)
Uninterrupted sine waves, usually of the RF type, that are radiated into space by a transmitter.

piezoelectric effect
The property of a crystal to change mechanical vibrations into electrical energy.

sharp filter
A filter that permits feedback only of the desired frequency.

pulse repetition rate (PRR)
The time that it takes for the waveform to repeat itself.

multivibrator
A type of relaxation oscillator that employs an RC network in its physical makeup and produces a rectangular-shaped wave.

triggered multivibrator
A type of multivibrator that uses a control technique called triggering to change its operational state.

free-running multivibrator
A type of multivibrator that is self-starting. It operates continuously as long as electrical power is supplied. The shape and frequency of the waveform is determined by component selection.

astable multivibrator
A free-running generator that develops a continuous square-wave output.

symmetrical multivibrator
A circuit with an output having equal on and off times.

nonsymmetrical multivibrator
A circuit with an output having unequal on and off times.

monostable multivibrator
A multivibrator with one stable state. It changes to the other state momentarily and then returns to its stable state.

flip-flop
A type of multivibrator changes states only when a trigger pulse is applied. It is called a flip-flop because it flips to one state when triggered and flops back to the other state when triggered.

threshold terminal
The beginning or entering point of an operating condition. A terminal connection of the LM555 IC timer.

blocking oscillator
An oscillator circuit that drives an active device into cutoff or blocks its conduction for a certain period of time.

vertical blocking oscillator (VBO)
A TV circuit that generates the vertical sweep signal for deflection of the cathode-ray tube or picture tube.

Chapter 21: Radio Frequency (RF) Communications Systems – Chapter Outline

Introduction
Objectives
Key Terms
21.1 RF Communications Systems
  RF Communication System Parts
  Self-Examination
21.2 Continuous Wave Communication
  Basic CW Parts and Operation CW Transmitter
  CW Receiver
  Antenna
  Signal Selection
  RF Amplification
  Heterodyne Detection
  Beat Frequency Oscillator
  AF Amplification
Self-Examination
21.3 AM Communications
  Modulation
  Percentage of Modulation
  AM Communication Systems
  AM Transmitter
  Simple AM Receiver
  Superheterodyne Receiver
  Self-Examination
21.4 FM Communications

Chapter 21 Objectives

After studying this chapter, you will be able to:
21.1 Explain the basic operation of an RF communication system.
21.2 Analyze the operation of a continuous-wave communication system.
21.3 Analyze the operation of an AM communication system.
21.4 Analyze the operation of an FM communication system.
21.5 Troubleshoot an RF communication circuit.

Chapter 21 Key Terms

beat frequency oscillator (BFO)
buffer amplifier
carrier wave
center frequency
channel
ganged
ground wave
heterodyning
high-level modulation
ionosphere
line-of-sight transmission
low-level modulation
modulating component
modulation
phasor

radio telegraphy
ratio detector
receiver
sidebands
sky wave
transmitter
zero beating

Chapter 21 – Radio Frequency (RF) Communications – Figure List

Figure 21-1. Electromagnetic Spectrum.
Figure 21-2. RF communication system with Transmitter-receiver.
Figure 21-3. Electromagnetic field changes. .
Figure 21-4. Sky-wave and ground-wave patterns.
Figure 21-5. Morse code.
Figure 21-6. Comparison of CW signals. A−CW signal. B−Coded or keyed continuous wave, PlaceNameletter U.
Figure 21-7. Block diagram of a CW communication system.
Figure 21-8. Single-stage CW transmitter.
Figure 21-9. Master Oscillator Power Amplifier (MOPA) transmitter.
Figure 21-10. CW receiver functions. . 16-10, EE5
Figure 21-11. Antenna signal voltage and current. . 16-11, EE5
Figure 21-12. Input tuner of a CW receiver and its bandwidth A−Input tuning circuit. B−Tuner frequency response curve. . 16-12, EE5
Figure 21-13. Antenna coils. . 16-13, EE5
Figure 21-14. Tuned radio frequency receiver. . 16-14, EE5
Figure 21-15. Heterodyne detector frequencies. . 16-15, EE5
Figure 21-16. Heterodyne diode output. . 16-16, EE5
Figure 21-17. Hartley oscillator used as a Beat Frequency Oscillator (BFO). . 16-17, EE5
Figure 21-18. AM signal components. . 16-18, EE5
Figure 21-19. Sideband frequencies of an AM signal. . 16-19, EE5
Figure 21-20. AM modulation levels. A−Under modulated signal. B−Fully modulated signal. C−Over modulated signal. . 16-20, EE5
Figure 21-21. AM communication system. . 16-21, EE5
Figure 21-22. Simplified AM transmitter. . 16-22, EE5
Figure 21-23. 5-W AM transmitter. . 16-23, EE5

Chapter 21 Summary

- An RF communication system consists of a transmitter and one or more receivers.
- The transmitter of an RF communication system serves as the signal source.
- The receiver of an RF communication system detects, amplifies, and reproduces the signal.
- Intelligence is applied to the transmitter according to the design of the communications system.
- Three types of communication systems are continuous wave (CW), amplitude modulation (AM), and frequency modulation (FM).
- Electromagnetic waves radiate from the antenna of the transmitter.
- Low-frequency waves (30 kHz to 300 kHz) and medium frequency waves (300 kHz to 3000kHz) tend to follow the curvature of the earth; this type of radiation pattern is called *ground wave*.
- Very high-frequency waves (30 MHz to 300 MHz) tend to move in a straight line; this type of radiation pattern is called *line-of-sight transmission*.
- A continuous wave (CW) communication system generates and electromagnetic wave of a constant frequency and amplitude.

- An amplitude modulation (AM) communication system changes the amplitude of a generated RF signal by the intelligence signal.
- A frequency modulation (FM) communication system superimposes intelligence on the carrier by variations in frequency.
- A simple CW communication system has an oscillator, power supply, and antenna; the receiver has an antenna-ground network, a tuning circuit with an RF amplifier, a heterodyne detector, a beat-frequency oscillator, an audio amplifier, and a speaker.
- To overcome the shortcomings of a single-stage CW transmitter, a power amplifier is incorporated after the oscillator.
- The signal selection function of radio receivers refers to its ability to pick out a desired RF signal.
- The purpose of heterodyning is to mix two AC signals.
- The space that an RF signal occupies along the frequency spectrum is called a channel.
- A radio receiver is responsible for signal interception, selection, demodulation and reproduction.
- A ratio detector is used as an FM demodulator circuit.

Chapter 21 Formulas

(21-1) $B_1 = F_1 + F_2$ Beat frequency 1.

(21-2) $B_2 = F_2 - F_1$ Beat frequency 2.

(21-3) % modulation = $\frac{V_{\max} - V_{\min}}{2V_{carrier}} \times 100$ Percentage of modulation in an AM communication system.

Chapter 21 Answers

Examples

21-1. 3.04 MHz

21-2. 25 MHz

21-3. 2 kHz

21-4. 125% (with $V_{min}=0V$)

21-5. 1005 kHz CW

21-6. Sum beat frequency = 1655 kHz AM, Difference beat frequency = 455 kHz AM. It may be interesting to note that the difference beat frequency is always 455 kHz AM in the IF band.

Self-Examination

21.1

1. transmitter
2. receiver
3. one-way
4. two-way
5. continuous-wave (CW), amplitude modulation (AM), frequency modulation (FM) (any order)
6. sky

21.2

7. Morse
8. signal selection
9. heterodyning
10. mixer
11. power supply, code key, RF oscillator, RF power amplifier and antenna
12. By interrupting the RF wave produced by the oscillator
13. Constant amplitude RF signal
14. This is the 0 beating condition and no sound output will occur
15. A low tone AF is produced

21.3

16. modulation
17. 535, 1620
18. undermodulation, overmodulation, 100% modulation (any order)
19. oscillator, AF component, antenna, power supply (any order)
20. antenna, RF tuner, demodulator, sound output (any order)
21. diode *or* half-wave rectifier
22. superheterodyne
23. intermediate frequency *or* IF
24. 50%
25. The diode works as a half wave rectifier of the incoming AM signal. This half-wave rectified voltage is used to charge a capacitor at an AF rate, thereby extracting the audio portion of the signal.

21.4

26. 88, 108
27. center frequency
28. 10.7
29. ratio

30. It detects the phase relationship of an FM signal applied to its input.
31. The manner in which the carrier signal changes with respect to the amplitude of the audio frequency signal determines the type of modulation. If the modulating component causes a change in the amplitude of the carrier it is called amplitude modulation, and if it causes a change is in the frequency of the carrier it is called frequency modulation.
32. amplitude, phase relationship

Chapter 21 Glossary

transmitter
The signal source of an RF communications system.

receiver
The component of an RF communications system that picks a signal out of the atmosphere and uses it to reproduce sound.

ground wave
Electromagnetic waves that follow the curvature of the earth.

sky wave
An RF signal radiated from an antenna into the ionosphere.

ionosphere
A layer of ionized particles in the atmosphere.

line-of-sight
An RF signal transmission that radiates out in straight lines because of its short wavelength.

radio telegraphy
The process of conveying messages by coded telegraph signals.

ganged
Two or more components connected together by a common shaft-a three-ganged variable capacitor.

heterodyning
The process of combining signals of independent frequencies to obtain a different frequency.

beat frequency
A resulting frequency that develops when two frequencies are combined in a nonlinear device.

beat frequency oscillator (BFO)
An oscillator of a CW receiver. Its output beats with the incoming CW signal to produce an audio signal.

zero beating
The resulting difference in frequency that occurs when two signals of the same frequency are heterodyned.

modulation
The process of changing some characteristic of an RF carrier so that intelligence can be transmitted modulating component.

demodulation
The process of extracting the low frequency intelligence (usually audio) signal form the high frequency carrier signal.

carrier wave
An RF wave to which modulation is applied.

sidebands
The frequencies above and below the carrier frequency that are developed because of modulation.

amplitude modulation
The communication process in which the amplitude of the carrier wave is varied according to the changes in the amplitude or frequency of the low frequency (usually audio) component.

frequency modulation
The communication process in which the frequency of the carrier wave is varied according to the changes in the amplitude or frequency of the low frequency (usually audio) component.

channel
The space that an AM signal occupies with its frequency

continuous wave
An electromagnetic wave of a constant frequency and amplitude. Intelligence is injected into the continuous wave by interrupting it with a coded signal.

channel
The space that an AM signal occupies with its frequency

Superheterodyne
A communicaiton receiving circuit that uses a local oscillator to produce modulated IF signal for accomplishings basic receiver functions

high-level modulation
A situation in which the modulating component is added to an RF carrier in the final power output of the transmitter.

antenna
A component responsible for transmitting and receiving radio frequency signals.

low-level modulation
A situation in which the modulating component is added to an RF carrier in the RF oscillator circuit.

buffer amplifier
An RF amplifier that follows the oscillator of a transmitter. It isolates the oscillator from the load.

center frequency
The carrier wave of an FM system without modulation applied.

ratio detector
A circuit used as a demodulator in FM radio receivers.

phasor
A line used to denote value by its length and phase by its position in a vector diagram.

Chapter 22: Specialized Communications Systems – Chapter Outline

Introduction
Objectives
Key Terms
22.1 Television Communication Systems
    Television Transmitter
    Television Receiver
    Color Television
    Color Camera
    Color Transmitter
    Color Receiver
22.2 Color Displays
    The Color CRT
    LCD (Liquid Crystal Display)

Chapter 22 Objectives

After studying this chapter, you will be able to:

22.1 Describe the basic operation of a television communications system
22.2 Identify and explain the technologies used in digital communication systems

Chapter 22 Key Terms

Analog to Digital Converter (ADC)
Digital to Analog Converter (DAC)
Access Point (AP)
Display unit
Deflection yoke
Pixel
Synchronization
Modulation
Charge Coupled Device (CCD)
Frame
Interlaced scanning
Progressive scanning
Intensity
CityplaceHue
Cathode Ray Tube (CRT)
Liquid Crystal Display (LCD)
Digital television (DTV)

High Definition Television (HDTV)
Luminance (Y signal)
Plasma display
Spread Spectrum
Digital Signal Processing (DSP)
Wireless Local Area Network (WLAN)

Chapter 22 – Communication Systems – Figure List

Figure 22-1. TV communication system: (a) transmitter; (b) receiver.
Figure 22-2. Simplification of a TV camera tube.
Figure 22-3. Scanning and sweep signals; (a) horizontal; (b) vertical.
Figure 22-4. Composite TV picture signal.
Figure 22-5. Monochrome TV transmitter.
Figure 22-6. Monochrome TV receiver.
Figure 22-7. Cathode-ray tube.
Figure 22-8. Simplified color TV camera. 16-44, EE5
Figure 22-9. Color TV transmitter block diagram. 16-45, EE5
Figure 22-10. Color TV receiver block diagram. 16-46, EE5
Figure 22-11. Cross-sectional view of a color television CRT display. pdf
Figure 22-12. Typical LCD display. pdf
Figure 22-13. Plasma screen technology – one pixel. pdf
*Figure 22-14. Block diagram of a digital cell phone. NEW

Chapter 22 Summary

- Audio and video signals which can change in a continuous manner are termed as analog signals
- Signals which carry text and data which change only in specific or discrete intervals are termed as digital signals.
- Television is a specialized type of communication, that uses both AM and FM transmissions, requiring carriers for both the picture (AM) and the sound (FM).
- A microphone typically provides the audio input, and a television camera provides the video input in a TV transmitter.
- The TV receiver demodulates the picture and sound signals from the respective carrier waves, following which the video output is reproduced on a display and sound on a speaker.

- The scene being televised by a camera tube must be broken into very small parts called picture elements or pixels, which are scanned in a conventional TV camera using an electron beam, and in a CCD camera the voltage developed on the pixels is transferred out for processing.
- Picture production in a conventional TV system generally employs interlace scanning, where for producing one frame, the odd-numbered-line of the 535 horizontal lines in a frame is scanned first, followed by the even-numbered-lines.
- Digital TV consists of 18 different standards, which are a combination of the resolution, aspect ratio and frame rate.
- The highest resolution of a digital TV signal is generated by High Definition TV (HDTV), which has a resolution of 1920 x 1080, having an aspect ratio of 16:9.
- In television signal production, the vertical and horizontal sweep circuits must be properly synchronized to produce a picture.
- In a composite TV picture signal, serrated pulses provide continuous horizontal sync during the vertical retrace time, and the vertical sync pulse is made up of six rectangular pulses near the center of the vertical blanking time.
- The FCC (Federal Communications Commission) has allocated a 6-MHz bandwidth for each TV channel.
- The brightness or luminance signal, referred to as the Y signal, is a mixture of the three primary colors, in the proportion 59% green, 30% red, and 11% blue.
- Color transmitters combine the three primary color signals into an I signal and a Q signal. The I signal refers to orange or cyan signals, and the Q signal corresponds to green or purple information. Both signals are used to amplitude modulate a 3.58-MHz subcarrier signal.
- The human eye needs stimuli in the form of hue and saturation stimuli to perceive color. Hue refers to a specific color, and saturation to the amount, or level, of color present.
- Color receivers, after demodulation, divide the composite video signal into the Y signal and the chroma (C) signal which in turn contains the I and Q color, and a 3.58-MHz suppressed carrier signal.
- A liquid crystal display (LCD), consists of liquid crystals sensitive to electrical current and capable of polarizing light placed between transparent polarizing filters and glass plates.
- Plasma displays are based on the illumination of a phosphor coated surface by ionized gas or plasma.

- The coverage area of cellular phone towers is called a cell.
- In a typical cellular phone communication system, the transmitter contains an Analog to Digital Converter (ADC), which converts the analog audio input signal into a digital format.
- Data in digital format can be suitably transformed for used in communication systems using Digital Signal Processing (DSP) methods.
- Modulation is used for changing some property of a high frequency carrier using the information that is to be transmitted, and demodulation extracts the information from the carrier.
- In a typical communication system, the receiver contains a Digital to Analog Converter (DAC) which converts the digital input to an analog audio output.
- The transmitter function of a cell phone is responsible for radiating a signal that has been developed by the microphone, and the receiver function is primarily responsible for detecting a RF signal received by the antenna and reproducing the audio component of this signal.
- A Wireless LAN (WLAN), uses RF signals for transmitting and receiving data between communicating devices.
- The base station used to extend a wireless LAN, which has the capability of communicating with wireless devices using an antenna is called an Access Point (AP).
- The range of coverage in a wireless LAN is reduced due to absorption of the RF signal by objects such as walls; by RF interference due to cordless phones; and by electrical interference due to motors.

Chapter 22 Answers

Self-Examination

22.1

1. 525
2. Amplitude
3. Frequency
4. red, blue, green
5. 16:9
6. 6
7. Progressive
8. video
9. plasma

10. CRT
11. VHF
12. I, Q
13. Diplexer
14. Demodulator
15. pixels

22.2

16. base unit, handset
17. 900 MHz, 2.4 GHz, 5.8 GHz
18. security
19. duplex
20. simplex
21. transmitting, receiving
22. Analog to Digital Converter (ADC)
23. Digital to Analog Converter (DAC)
24. microphone, speaker
25. roaming
26. after
27. before
28. demodulator
29. receiver
30. transmitter
31. 2.4 GHz, 5 GHz
32. Access Point
33. 800 MHz, 1800 MHz
34. 38m
35. 2.4 GHz

Chapter 22 Glossary

Access Point. The base station used to extend a wireless LAN, which has the capability of communicating with wireless devices using an antenna.

ADC. Analog to digital converter, a device which converts an analog signal into a digital format.

Ad hoc topology. An unmanaged wireless network topology wherein devices can communicate directly with each other.

Aspect ratio. The ratio of the unit width to the unit height of a display device.

Blanking pulse. A part of the TV signal where the electron beam is turned off during the retrace period. There is both vertical and horizontal blanking.

Channel. The band of frequencies used for wireless communication

Chroma. Short for chrominance. Refers to color in general.

Compatible. A TV system characteristic in which broadcasts in color may be received in black and white on sets not adapted for color.

DAC. Digital to Analog converter, a device which converts a digital signal into an analog one.

Deflection. Electron beam movement of a TV system that scans the camera tube or picture tube.

Deflection yoke. A coil fixture that moves an electron beam vertically and horizontally.

DSP. Digital Signal Processing of information which is in digital format.

Directional antenna. An antenna which is focused at a receiver for concentrating the signal power in a certain direction.

Diplexer. A special TV transmitter coupling device that isolates the audio carrier and the picture carrier signals from each other.

Duplex. The ability of a system to transmit and receive signals from the same unit.

Encryption. The process of altering or transforming data, in order to prevent unauthorized access.

Field, even/odd lined. The even/odd-numbered scanning lines of one TV picture or frame.

Frame. A complete electronically produced TV picture of 525 horizontally scanned lines.

CityplaceHue. A color, such as red, green, or blue.

Infrastructure topology. A managed wireless network topology in which a wired LAN is extended by using a wireless base station, in order to support wireless devices.

Interlace scanning. An electronic picture production process in which the odd numbered lines of a display are scanned first, and then the even lines are scanned next, and interleaved in order to make a complete 525-line picture.

Intermediate frequency (IF). A single frequency that is developed by heterodyning two input signals together in a superheterodyne receiver.

I-signal (1). A color signal of a TV system that is in phase with the 3.58-MHz color subcarrier.

LCD. A liquid crystal display, which consists of liquid crystals sensitive to electrical current and capable of polarizing light placed between transparent polarizing filters and glass plates.

LAN. A Local Area Network or computer network operating within a relatively small geographic area such as a home or a small office

Monochrome. Black-and-white television.

Negative picture phase. A video signal characteristic where the darkest part of a picture causes the greatest change in signal amplitude.

Omni-directional antenna. An antenna used for transmitting signals in all directions.

Pixel. A discrete picture element or photosensitive area of a display.

Plasma display. A display which utilizes an ionized gas to activate a picture element.

Progressive scanning. An electronic picture production process in which all the lines of a display are scanned consecutively to make a complete 525-line picture.

Protocol. In computer based communication systems, a set of rules used by devices for exchanging information over a given media.

Q signal. A color signal of a TV system that is out of phase with the 3.58-MHz color subcarrier.

Video Resolution. The process of separating or distinguishing between detailed picture elements and is specified by number of pixels used to produce a display.

Retrace. The process of returning the scanning beam of a camera or picture tube to its starting position.

Saturation. The strength or intensity of a color used in a TV system.

Scanning. In a TV system, the process of moving an electron beam vertically and horizontally.

Spread Spectrum. The manner in which data signals travel through the wireless media either altering between frequency carriers or spreading the signal over the frequency spectrum.

Sync. An abbreviation for synchronization.

Synchronization. A control process that keeps electronic signals in step. The sweep of a TV receiver is synchronized by the transmitted picture signal.

Topology. The layout of the computer network, for transmitting and receiving data, including the physical cabling and logical flow of information.

Trace time. A period of the scanning process where picture information is reproduced or developed.

Transceiver. A communication device which is used both for transmitting and receiving signals.

WLAN. A Wireless LAN, which uses RF signals for transmitting and receiving data between communicating devices.

Y signal. The brightness or luminance signal of a TV system.

Chapter 23: Digital Electronic Systems – Chapter Outline

Introduction
Objectives
Key Terms
23.1 Digital Systems
    Decimal Number Systems
    Binary Number Systems
    Binary Fractions
    Binary-Coded Decimal Numbers
    Octal Numbering Systems
    Hexadecimal Numbering Systems
    Self-Examination
23.2 Digital Logic Circuits
    Binary Logic Functions
    AND Gates
    OR Gates
    Combinational Logic Gates
    Self-Examination
23.3 Flip-Flops

Chapter 23 Objectives

After studying this chapter, you will be able to:

1. Explain the electrical differences in the response of a system that is used for analog or digital applications.
2. Explain the differences between analog and digital systems.
3. Change decimal numbers to an equivalent binary, binary coded decimal (BCD), octal, or hexadecimal numbers.
4. Change binary, BCD, or hexadecimal numbers to equivalent decimal numbers.
5. Identify basic logic symbols of AND, OR, NOT, NOR, and NAND gates.
6. Analyze and develop logic equations and truth tables for logic gates.
7. Evaluate the operation of transistor logic gates.
8. Describe the operation of *RS, D,* and JK flip-flops (triggered and clocked).
9. Describe the operation of counting circuits.

Chapter 23 Key Terms

Amplification
Analog
Base
Binary
Decoding
Digital
Digital integrated circuit
Dual-in-line package (DIP
Encoding
Energy
Gate
Hexadecimal
Logic

Octal
Pulse
Radix
Semiconductor
Solid state
System
Transistor
AND gate
Bistable
Boolean algebra
NAND gate
NOT gate
OR gate
Truth table
Asynchronous
Bistable
Clear
Clock input
Debounce
Flip-flop
Latch
Memory
Register
Reset
Toggle
BCD counter
Binary counter
Decade counter
Down counter
Incrementing
Modulo
Register
Shift register
Up counter

Chapter 23 – Digital Electronic Systems – Figure List

Chapter 23 Summary

- A digital system has a source of operating energy, a path, control, a load device, and an indicator.
- A decimal numbering system has 10 individual values or symbols.
- Nearly all digital systems use binary numbers in their operation.
- A binary number has 2 as its base and uses only 1s and 0s and place values are expressed as powers of 2.
- Conversion of a decimal number to a binary number is achieved by repetitive steps of division by the number 2.
- A binary-coded decimal number is used to indicate large binary numbers.
- Octal, or base 8, numbering systems are used to process large numbers through a digital system.
- Hexadecimal numbers used in digital systems have a base of 16 with digits 0 through 9 and the letters A, B, C, D, E, and F.

- Binary codes have been developed to interface a digital system between binary data and alphanumeric data.
- The BCD code was developed to display decimal numbers in groups of four binary bits.
- The parity code is an error-checking code.
- The odd parity code generates a 0 parity bit when the number of ls is odd.
- The even parity code is just the opposite of the odd code.
- Three basic gates have been developed to make logic decisions– AND, OR, and NOT.
- An AND gate is designed to have two or more inputs and one output, if all inputs are 1, then the output will be 1. If any input is 0, then the output will be 0.
- Mathematically, the action of an AND gate is expressed as $A \times B = C$. This is the multiplication operation.
- An OR gate is designed to have two or more inputs and a single output, so an OR gate will produce a 1 output when a 1 appears at any input.
- Mathematically, the OR gate function is expressed as $A + B = C$, called OR addition.
- A NOT gate has a single input and a single output and is achieved by an inverter with a 1 input causing the output to be 0 and a 0 input causing the output to be 1.
- Mathematically, the operation of a NOT gate is expressed as $A = A'$ (inverse)..
- When two of the basic logic gates are connected together, they form a combination logic gate., with the two most common the NOT-AND and NOT-OR, referred to as NAND and NOR gates..
- A NAND gate is an inversion of the AND function, so when a 1 appears at all inputs the output will be 0 and when a 0 appears at any input the output will be 1.
- Mathematically, the operation of a NAND gate is expressed as $A \times B = C$.
- A NOR gate produces an inversion of the OR function, so when the inputs are all 0, the output is 1 and when any input is 1, the corresponding output is 0.
- Mathematically, the operation of a NOR gate is expressed as $A + B = C$.
- A flip-flop is a logic device with two or more inputs and two outputs.
- A bistable latch has cross-connected NAND or NOR gates.
- An RS flip-flop can be designed from two cross-coupled NAND or NOR gates similar to the bistable latch.

- The clocked signal applied to a flip-flop is generally a rectangular-shaped series of waves and are defined as level triggering and edge triggering.
- In level triggering, the state of a clock changes value from 0 to 1 and carries out a transfer of data.
- Positive-edge triggering occurs at the leading edge of a pulse.
- Negative-edge triggering occurs at the trailing edge of a pulse.
- An edge-triggered flip-flop can have its data changed anytime, whereas level triggering can be changed only once at a specific level.
- A clocked RS flip-flop has SET, RESET, and CLOCK inputs with Q and Q outputs, with the inputs either positive- or negative-edge triggered.
- The outputs of an RS flip-flop are identified as Q and Q' (inverse).
- A D flip-flop is a clocked RS flip-flop with an inverter connected across its inputs.
- Digital counters are made up of flip-flops.
- Three flip-flops grouped together form a binary-coded octal counter called a modulo-8 counter.
- A modulo 8 counter goes from 000 to 111_2 and represents 7_{10}.
- When four flip-flops are connected together in a group, it is possible to develop the units part of a binary-coded hexadecimal counter, called a modulo- 16 counter.
- Four flip-flops are commonly connected together to form a binary-coded decimal counter.
- Synchronous, or parallel, input counters trigger each flip-flop at the same time by the clock signal.

Chapter 23 Glossary

Analog
A quantity that is continuous or has a continuous range of values.

Base
The number of symbols in a number system. A decimal system has a base of 10, a binary system has a base of 2, an octal system has a base of 8, and a hexadecimal system has a base of 16.

Binary
A system of numerical representation that uses only two symbols, 1 and 0.

Decoding
A function of a digital system that is responsible for changing coded data into alphanumeric data.

Digital
A value or quantity related to numbers of discrete values.

Digital integrated circuit
An IC that responds to two states (on-off) data.

Dual-in-line package (DIP)
A packaging method for integrated circuits.

Encoding
A function of a digital system that is responsible for changing input data, which may be in analog form, into binary data.

Energy
Something that is capable of producing work, such as heat, light, chemical, and mechanical action.

Gate
A circuit that performs special logic operations such as AND, OR, NOT, NAND, and NOR.

Hexadecimal
A base 16 numbering system that uses the symbols 0, 1, 2, 3, 4, 5, 6, 7, 8, 9, A, *B, C, D, E,* and F

Integrated circuit (IC)
A circuit in which many elements are fabricated and interconnected by a single process, as opposed to a non-integrated circuit, in which individual components such as resistors, diodes, and transistors are fabricated and then assembled.

Logic
A decision-making capability of computer circuitry.

Memory
The storage capability of a device or circuit.

Octal
A base 8 numbering system that uses the symbols 0, 1, 2, 3, 4, 5, 6, and 7.

Pulse
A nonsinusodial signal or wave that occurs randomly or on a periodical basis that can be generated electronically and used to control a digital system.

Radix
The base of a numbering system.

Semiconductor
An element such as silicon or germanium that is intermediate in electrical conductivity between an insulator and a conductor.

Solid state
An area of electronics dealing with the conduction of current carriers through semiconductors.

System
A combination of functional parts (energy source, path, control, load, and indicator), which are needed to make a piece of equipment operate.

Transistor
A semiconductor device capable of transferring a signal from one circuit to another and producing amplification.

Bistable
Any device that can be set into one of two operational states or conditions such as on and off or 1 and 0.

Boolean algebra
Binary logical algebra that provides information in the form of equations and expressions.

NAND gate
A logic gate that produces a 1, or high output, for any input combination except when Is are applied to each input.

NOR gate
A logic gate that will produce a 1, or high output, only when the inputs are all 0s, or low.

NOT gate
An inverter that changes the polarity of the input in its output.

OR gate
A logic gate that provides a 1 output if there is a 1 applied to any of its inputs.

Truth table
A graph or table that displays the operation of a logic circuit with respect to its input and output data.

Asynchronous
Signals or events that can occur at any time without reference to a system clock.

Clear
An asynchronous input to a flip-flop; also called reset used to restore the Q output of a flip-flop to the logic 0 state.

Clock
A pulse generator that controls the timing of a computer through signals applied to various components.

Clock input
The terminal of a flip-flop whose condition changes with a clock signal to provide synchronized control.

Flip-flop
A digital circuit component having a stable operating condition and the capability of changing from one state to another with the application of a signal pulse.

Latch
A simple logic storage element such as a flip-flop that is used to retain data in an operational state.

Memory
A collection of latches that can store a number of different logic levels.

Register
A small group of latches used to store several bits of data.

Reset
A flip-flop input that achieves the same function as clear.

Synchronous
A digital circuit where all of the ICs are paced by a common clock signal and no activity occurs between clock pulses.

Set
An asynchronous input used to restore the Q output of a flip-flop to a logic 1 level (Some flip-flops use the term *preset* for this function).

Toggle
A condition describing the output of a flip-flop where the Q and Q' output of a flip-flop changes to the complement of its previous state on each transition of the clock.

BCD counter
A circuit that counts to 10 in binary-coded numbers. A binary-coded-decimal counter.

Binary counter
A counter that progresses through a straight binary counting sequence.

Counter
A device capable of changing states in a specified sequence upon receiving appropriate input data and producing an output that is indicative of the number of input pulses.

Decade counter
A circuit that counts through 10 distinct states. Decrementing. Decreasing the value of a counter by a known amount.

Incrementing
Increasing the value of a counter by a known amount.

Modulo
A value with respect to the modulus of a body or device such as a *modulo 5* counter that responds by making the equivalent of 5 counts.

Modulus
A value that expresses numerically the degree to which a property is possessed by a body or device.

Register
A group of flip-flops that can be used to store binary information.

Shift register
A collection of flip-flops connected so that a binary number can be shifted into or out of the flip-flops.

Up counter
A circuit that counts from its smallest numerical value to its largest value in binary progression and is incremental.

Milton Keynes UK
Ingram Content Group UK Ltd.
UKHW031151141024
449569UK00024B/876